WITH SPEED AND VIOLENCE

WITH SPEED AND VIOLENCE

Why Scientists Fear Tipping Points
in Climate Change

FRED PEARCE

BEACON PRESS
Boston

BEACON PRESS
25 Beacon Street
Boston, Massachusetts 02108-2892
www.beacon.org

Beacon Press books
are published under the auspices of
the Unitarian Universalist Association of Congregations.

10 09 08 07   8 7 6 5 4 3 2

This book is printed on acid-free paper that meets the uncoated paper
ANSI/NISO specifications for permanence as revised in 1992.

Composition by Wilsted & Taylor Publishing Services

Library of Congress Cataloging-in-Publication Data

Pearce, Fred.
With speed and violence : why scientists fear tipping points in climate change / Fred Pearce.
    p. cm.
Includes index.
ISBN-13: 978-0-8070-8576-9 (hardcover : alk. paper)
ISBN-10: 0-8070-8576-6 (hardcover : alk. paper) 1. Climatic changes. 2. Climatic changes—
History—Chronology. I. Title.

QC981.8.C5P415 2006
551.6—dc22   2006019901

We are on the precipice of
climate system tipping points
beyond which there is no redemption.

JAMES HANSEN, *director,*
NASA Goddard Institute for Space Studies,
New York, December 2005

CONTENTS

III. Riding the carbon cycle

IV. Reflecting on warming

V. Ice ages and solar pulses

VI. Tropical heat

VII. At the millennium

VIII. Inevitable surprises

CHRONOLOGY OF CLIMATE CHANGE

5 billion years ago Birth of planet Earth

600 million years ago Last occurrence of "Snowball Earth," followed by warm era

400 million years ago Start of long-term cooling

65 million years ago Short-term climate conflagration after meteorite hit

55 million years ago Methane "megafart" from ocean depths causes another short-term conflagration

50 million years ago Cooling continues as greenhouse-gas levels in air start to diminish

25 million years ago First modern ice sheet starts to form on Antarctica

3 million years ago First ice-sheet formation in the Arctic ushers in era of regular ice ages

100,000 years ago Start of most recent ice age

16,000 years ago Most recent ice age begins stuttering retreat

14,500 years ago Sudden warming causes sea levels to rise 65 feet in 400 years

12,800 years ago Last great "cold snap" of the ice age, known as the Younger Dryas era, is triggered by emptying glacial lake in North America and continues for around 1,300 years before ending very abruptly

8,200 years ago Abrupt and mysterious return to ice-age conditions for several hundred years, followed by warm and stable Holocene era

8,000 years ago Storegga landslip in North Sea, probably triggered by methane clathrate releases that also bolster the warm era

5,500 years ago Sudden aridification of the Sahara

4,200 years ago Another bout of aridification, concentrated in the Middle East, causes widespread collapse of civilizations

1,200 to 900 years ago Medieval warm period in the Northern Hemisphere; megadroughts in North America

700 to 150 years ago Little ice age in the Northern Hemisphere, peaking in the 1690s

1896 Svante Arrhenius calculates how rising carbon dioxide levels will raise global temperatures

1938 Guy Callendar provides first evidence of rising carbon dioxide levels in the atmosphere, but findings ignored

1958 Charles Keeling begins continuous monitoring program that reveals rapidly rising carbon dioxide levels in the atmosphere

1970s Beginning of strong global warming that has persisted ever since, almost certainly attributable to fast-rising carbon dioxide emissions, accompanied by shift in state of key climate oscillations such as El Niño and the Arctic Oscillation, and increased melting of the Greenland ice sheet

Early 1980s Shocking discovery of Antarctic ozone hole brings new fears of human influence on global atmosphere

1988 Global warming becomes a front-page issue after Jim Hansen's presentations in Washington, D.C., during U.S. heat wave

1992 Governments of the world attending Earth Summit promise to prevent "dangerous climate change" but fail to act decisively

1998 Warmest year on record, and probably for thousands of years, accompanied by strong El Niño and exceptionally "wild weather," especially in the tropics; major carbon releases from burning peat swamps in Borneo

2001 Government of Tuvalu, in the South Pacific, signs deal for New Zealand to take refugees as its islands disappear beneath rising sea levels

2003 European heat wave—later described as the first extreme-weather event attributable to man-made global warming—kills more than 30,000; a third of the world is reported as being at risk of drought: twice as much as in the 1970s

2005 Evidence of potential "positive feedbacks" accumulates with exceptional hurricane season in the Atlantic, reports of melting Siberian permafrost, possible slowing of ocean conveyor, escalating loss of Arctic sea ice, and faster glacial flow on Greenland

THE CAST

Richard Alley, Penn State University, Pennsylvania. A glaciologist and leading analyst of Greenland ice cores, Alley is one of the most articulate interpreters of climate science. He has revealed that huge global climate changes have occurred over less than a decade in the past.

Svante Arrhenius, a Swedish chemist. In the 1890s, he was the first to calculate the likely climatic impact of rising concentrations of carbon dioxide in the atmosphere, and thus invented the notion of "global warming." Modern supercomputers have barely improved on his original calculation.

Gerard Bond, formerly of Lamont-Doherty Earth Observatory, Columbia University, New York. A geologist, Bond was one of the first analysts of deep-sea cores; until his death, in 2005, he was an advocate of the case that regular pulses in solar activity drive cycles of climate change on Earth, such as the little ice age and the medieval warm period.

Wally Broecker, Lamont-Doherty Earth Observatory, Columbia University. An oceanographer and one of the most influential and controversial U.S. climate scientists for half a century, Broecker discovered the ocean conveyor, a thousand-year global circulation system that begins off Greenland and ends in the Gulf Stream, which keeps Europe warm.

Peter Cox, UK Centre for Hydrology and Ecology, Wareham. Cox is an innovative young climate modeler of aerosols' likely role in keeping the planet cool—and of the risks that land plants will turn from a "sink" for to a "source" of carbon dioxide later in this century.

James Croll, a nineteenth-century Scottish artisan and self-taught academic. After many years of study, he uncovered the astronomical causes of the ice ages, a discovery that was later attributed to the Serbian mathematician Milutin Milankovitch.

Paul Crutzen, Max Planck Institute for Chemistry, Mainz, Germany. An atmospheric chemist who won the Nobel Prize in 1995 for his work predicting the destruction of the ozone layer, Crutzen pioneered thinking about stratospheric chemistry, the role of man-made aerosols in shading the planet, and "nuclear winter," and coined the term "Anthropocene."

Joe Farman, formerly of the British Antarctic Survey, Cambridge. Farman's dogged collection of seemingly useless data was rewarded by discovery of the ozone hole over Antarctica.

Jim Hansen, director of NASA's Goddard Institute for Space Studies, New York. Hansen's unimpeachable scientific credentials have preserved his position as President George W. Bush's top climate modeler (as this book goes to press), despite his outspoken warnings that the world is close to dangerous climate change, which have clearly irked the Bush administration.

Charles David Keeling, formerly of Scripps Institution of Oceanography, La Jolla, California. Until his death, in 2005, Keeling had made continuous measurements of atmospheric carbon dioxide on top of Mauna Loa, in Hawaii, since 1958. The resulting "Keeling curve," the most famous graph in climate science, shows a steady annual rise superimposed on a seasonal cycle as Earth " breathes."

Sergei Kirpotin, Tomsk State University, Russia. Kirpotin is the ecologist who told the world about the "meltdown" of permafrost in the West Siberian peat lands, raising fears that massive amounts of methane would be released into the atmosphere.

Michael Mann, director of the Earth System Science Center, Penn State University, Pennsylvania. A climate modeler and the creator of the "hockey stick" graph, a reconstruction of past temperatures showing that recent warming is unique to the past two millennia, Mann is the butt of criticism from climate skeptics, but gives as good as he gets. He is the co-founder of the RealClimate Web site.

Peter deMenocal, Lamont-Doherty Earth Observatory, Columbia University, New York. A climate historian, deMenocal has charted mega-droughts, the sudden drying of the Sahara, and other major climate shifts of the past 10,000 years, and their role in the collapse of ancient cultures.

John Mercer, formerly of Ohio State University, Columbus. The glaciologist who first proposed that the West Antarctic ice sheet has an Achilles heel, and that a "major disaster" there may be imminent, Mercer also pioneered research on tropical glaciers.

Drew Shindell, NASA's Goddard Institute for Space Studies, New York. An ozone-layer expert and climate modeler, Shindell is doing groundbreaking research on unexpected links between the upper and the lower atmosphere, revealing how the stratosphere can amplify small changes in surface temperature.

Lonnie Thompson, Byrd Polar Research Institute, Ohio State University, Columbus. A geologist, Thompson has probably spent more time above 20,000 feet than any lowlander alive, all in the pursuit of ice cores from tropical glaciers that are rewriting the planet's climate history.

Peter Wadhams, head of polar ocean physics at the University of Cambridge. He rode in British military submarines to provide the first data on thinning Arctic sea ice and discovered the mysterious "chimneys" off Greenland where the global ocean conveyor starts.

PREFACE: THE CHIMNEY

The Greenland Sea occupies a basin between Greenland, Norway, Iceland, and the Arctic islands of Svalbard. It is like an antechamber between the Atlantic and the Arctic Ocean: the place where Arctic ice flowing south meets the warm tropical waters of the Gulf Stream heading north. Two hundred years ago, the sea was a magnet for sailors intent on making their fortunes by harpooning its great schools of bowhead whales. For a few decades, men such as the Yorkshire whaling captain and amateur Arctic scientist William Scoresby sailed north each spring as the ice broke up and dodged the ice floes to hunt the whales that had congregated to devour the spring burst of plankton. Scoresby was the star of the ice floes, landing a world-record thirty-six whales at Whitby Harbour after one trip in 1798. He was the nimblest navigator around a great ice spur in the sea known as the Odden tongue, where the whales gathered.

Scoresby was too clever for his own good, and boom turned to bust when all the whales had been killed. What was once the world's most prolific and profitable whaling ground is still empty of bowheads. But just as the unique mix of warm tropical waters and Arctic ice was the key to the Greenland Sea's whaling bonanza, so it is the key to another hidden secret of these distant waters.

It's called "the chimney." Only a handful of people have ever seen it. It is a giant whirlpool in the ocean, 6 miles in diameter, constantly circling counterclockwise and siphoning water from the surface to the seabed 2 miles below. That water will not return to the surface for a thousand years. The chimney, once one of a family, pursues its lonely task in the middle of one of the coldest and most remote seas on Earth. And its swirling waters may be the switch that can turn the heat engine of the world's climate system on and off. If anything could trigger the climatic conflagration shown in the Hollywood movie *The Day After Tomorrow*, it would be the chimney.

The existence of a series of these chimneys was discovered by a second British adventurer, Cambridge ocean physicist Peter Wadhams. In the 1990s, he began hitching rides in Royal Navy submarines beneath the Arctic ice. Like Scoresby, he was fascinated by his journeys to the Odden tongue—not for its long-departed whales, but because of the bizarre giant whirlpools he found there. He concluded that they were the final destination for the most northerly flow of the Gulf Stream. The waters of this great ocean current, which drives north through the tropical Atlantic bringing warmth to Europe, are chilled by the Arctic winds in the Greenland Sea and start to freeze around the Odden tongue. The water that is left becomes ever denser and heavier until it is entrained by the chimneys and plunges to the ocean floor.

This was a dramatic discovery. The chimneys were, Wadhams realized, the critical starting point of a global ocean circulation system that oceanographers had long hypothesized but had never seen in action. It traveled the world's oceans, passing south of Africa, around Antarctica, and through the Indian and Pacific Oceans, before gradually resurfacing and sniffing the air again as it returned to the Atlantic, joined the Gulf Stream, and moved north once again to complete a circulation dubbed by oceanographers the "ocean conveyor."

But even as he gazed on these dynamos of ocean circulation, Wadhams knew that they were in trouble. For the Arctic ice was disappearing. Sonar data he had collected from the naval submarines revealed that the entire ice sheet that once covered the Arctic was thinning and breaking up. By the end of the 1990s, the Odden tongue was gone. The Gulf Stream water still came north, but it never again got cold enough to form ice. The ice tongue has not returned.

"In 1997, the last year that the Odden tongue formed, we found four chimneys in a single season, and calculate there could have been as many as twelve," says Wadhams. Since then, they have been disappearing one by one—except for one particularly vigorous specimen. Wadhams first spotted it out in the open ocean, at 75° north and right on the Greenwich Mean Line, during a ship cruise in March 2001. By rights, it should not have been there without the ice, he says. But it was, hanging in there, propelled downward perhaps by the saltiness created by evaporation of the water in the wind.

He found the same chimney again later that summer, twice the following year, and a final time in spring 2003, before the British government cut off his research funds. Over the two years he tracked it, the last great chimney had moved only about 20 miles across the ocean, like an underwater tornado that refused to go away. Wadhams measured it and probed it. He sent submersible instruments down through it to measure its motion at depth. It rotated, he said, right to the ocean floor, and such was the force of the downward motion that it could push aside a column of water half a mile high. "It is amazing that it could last for more than a few days," Wadhams says. "The physics of how it did it is not understood at all."

The great chimney had in May 2003 one dying companion, 40 miles to the northwest. But that chimney no longer reached the surface and was, he says, almost certainly in its death throes. That left just one remaining chimney in the Greenland Sea. "It may be many decades old or just a transitory phenomenon," he says. "But either way, it, too, may be gone by now. We just don't know." Like Scoresby's bowheads, it may disappear unnoticed by the outside world. Or we may come to rue its passing.

INTRODUCTION

Some environmental stories don't add up. I'm an environment journalist, and sometimes the harder you look at a new scare story, the less scary it looks. The science is flaky, or someone has recklessly extrapolated from a small local event to create a global catastrophe. Ask questions, or go and look for yourself, and the story dissolves before your eyes. I like to question everything. I am, I hope in the best sense, a skeptical environmentalist. Sometimes it is bad for business. I have made enemies by questioning theories about advancing deserts, by pointing out that Africa may have more trees than it did a century ago, and by condemning the politics of demographic doomsday merchants.

But climate change is different. I have been on this beat for eighteen years now. The more I learn, the more I go and see for myself, and the more I question scientists, the more scared I get. Because this story does add up, and its message is that we are interfering with the fundamental processes that make Earth habitable. It is our own survival that is now at stake, not that of a cuddly animal or a natural habitat.

Don't take my word for it. Often in environmental science it is the young, idealistic researchers who become the impassioned advocates. Here I find it is the people who have been in the field the longest—the researchers with the best reputations for doing good science, and the professors with the best CVs and longest lists of published papers—who are the most fearful, often talking in the most dramatic language. People like President George W. Bush's top climate modeler, Jim Hansen, the Nobel Prize–winner Paul Crutzen, and the late Charles Keeling, begetter of the Keeling curve of rising carbon dioxide levels in the atmosphere. They have seemed to me not so much old men in a hurry as old men desperate to impart their wisdom, and their sense that climate change is something special.

Nature is fragile, environmentalists often tell us. But the lesson of this book is that it is not so. The truth is far more worrying. Nature is strong and packs a serious counterpunch. Its revenge for man-made global warming will very probably unleash unstoppable planetary forces. And they will not be gradual. The history of our planet's climate shows that it does not do gradual change. Under pressure, whether from sunspots or orbital wobbles or the depredations of humans, it lurches—virtually overnight. We humans have spent 400 generations building our current civilization in an era of climatic stability—a long, generally balmy spring that has endured since the last ice age. But this tranquility looks like the exception rather than the rule in nature. And if its end is inevitable one day, we seem to be triggering its imminent and violent collapse. Our world may be blown away in the process.

The idea for this book came while I sat at a conference, organized by the British government in early 2005, on "dangerous climate change" and how to prevent it. The scientists began by adopting neutral language. They made a distinction between Type I climate change, which is gradual and follows the graphs developed by climate modelers for the UN's Intergovernmental Panel on Climate Change (IPCC), and Type II change, which is much more abrupt and results from the crossing of hidden "tipping points." It is not in the standard models. During discussions, this temperate language gave way. Type II climate change became, in the words of Chris Rapley, director of the British Antarctic Survey, the work of climatic "monsters" that were even now being woken.

Later in the year, Jim Hansen spoke in even starker terms at a meeting of the American Geophysical Union, saying: "We are on the precipice of climate system tipping points beyond which there is no redemption." The purpose of this book is to introduce Rapley's monsters and Hansen's tipping points and to ask the question, How much time have we got?

The monsters are not hard to find. As I was starting work on this book, scientists beat a path to my door to tell me about them. I had an e-mail out of the blue from a Siberian scientist alerting me to drastic environmental change in Siberia that could release billions of tons of greenhouse gases from the melting permafrost in the world's biggest bog. Glaciologists, who are more used to seeing things happen slowly, told me of dramatic events in Greenland and Antarctica, where they are discovering huge river systems of meltwater beneath the ice sheets, and of events in Pine Island Bay, one of the most remote spots in Antarctica, that they discussed with a shudder. Soon, they said, we could be measuring sea level rise in feet rather than inches.

Along the way, I also learned about solar pulses, about the "ocean conveyor," about how Indian village fires may be melting the Arctic, about a rare molecule that runs virtually the entire clean-up system for the planet, and above all about the speed and violence of past natural climate change. Some of this, I admit, has the feel of science fiction. On one plane journey, I reread John Wyndham's sci-fi classic *The Kraken Wakes,* and was struck by the similarities between events he describes and predictions for the collapse of the ice sheets of Greenland and Antarctica. It is hard to escape the sense that primeval forces lurk deep in the ocean, in ice caps, in rainforest soils, and in Arctic tundra. Hansen says that we may have only one decade, and one degree of warming, before the monsters are fully awake. The worst may not happen, of course. Nobody can yet prove that it will. But, as one leading climate scientist put it when I questioned his pessimism, how lucky do we feel?

I hope I have retained my skepticism through this journey. One of the starting points, in fact, was a reexamination of whether the climate skeptics—those who question the whole notion of climate change as a threat—might be right. Much of what they say is political hyperbole, of more benefit to their paymasters in the fossil-fuel lobby than to science. Few of them are climate scientists at all. But in some corners of the debate, they have done good service. They have, for instance, provided a useful corrective to the common assumption that all climate change must be man-made. But my conclusion from this is the opposite of theirs. Far from allowing us to stop worrying about man-made climate change, the uncertainties they highlight underline how fickle climate can be and how vulnerable we may be to its capricious changes. As Wally Broecker, one of the high priests of abrupt planetary processes, says, "Climate is an angry beast, and we are poking it with sticks."

This book is a reality check about the state of our planet. That state

scares me, just as it scares many of the scientists I have talked to—sober
scientists, with careers and reputations to defend, but also with hopes for
their own futures and those of their children, and fears that we are the last
generation to live with any kind of climatic stability. One told me quietly:
"If we are right, there are really dire times ahead. Having a daughter who
will be about my present age in 2050, and will be in the midst of it, makes
the issue more poignant."

I

WELCOME TO THE ANTHROPOCENE

I

THE PIONEERS

The men who measured the planet's breath

This story begins with a depressed Swedish chemist, alone in his study in the sunless Nordic winter after his marriage to his beautiful research assistant, Sofia, had collapsed. It was Christmas Eve. What would he do? Some might have gone out on the town and found themselves a new partner. Others would have given way to maudlin sentiment and probably a few glasses of beer. Svante Arrhenius chose neither release. Instead, on December 24, 1894, as the rest of his countrymen were celebrating, he rolled up his sleeves, settled down at his desk, and began a marathon of mathematical calculations that took him more than a year.

Arrhenius, then aged thirty-five, was an obdurate fellow, recently installed as a lecturer in Stockholm but already gaining a reputation for rubbing his colleagues the wrong way. As day-long darkness gave way to months of midnight sun, he labored on, filling book after book with calculations of the climatic impact of changing concentrations of certain heat-trapping gases on every part of the globe. "It is unbelievable that so trifling a matter has cost me a full year," he later confided to a friend. But with his wife gone, he had few distractions. And the calculations became an obsession.

What initially spurred his work was the urge to answer a popular riddle of the day: how the world cooled during the ice ages. Geologists knew by then that much of the Northern Hemisphere had for thousands of years been covered by sheets of ice. But there was huge debate about why this might have happened. Arrhenius reckoned that the clue lay in gases that could trap heat in the lower atmosphere, changing the atmosphere's radiation balance and altering temperatures.

He knew from work half a century before, by the French mathematician Jean Baptiste Fourier and an Irish physicist called John Tyndall, that some gases, including carbon dioxide, had this heat-trapping effect. Tyndall had measured the effect in his lab. Put simply, it worked like this: the gases were transparent to ultraviolet radiation from the sun, but they trapped the infrared heat that Earth's surface radiated as it was warmed by the sun. Arrhenius reasoned that if these heat-trapping gases in the air decreased for some reason, the world would grow colder. Later dubbed "greenhouse gases," because they seemed to work like the glass in a greenhouse, these gases acted as a kind of atmospheric thermostat.

Tyndall, one of the most famous scientists of his day and a friend of Charles Darwin's, had himself once noted that if heat-trapping gases were eliminated from the air for one night, "the warmth of our fields and gardens would pour itself unrequited into space, and the sun would rise upon an island held fast in the iron grip of frost." That sounded to Arrhenius very much like what had happened in the ice ages. Sure enough, when he emerged from his labors, he was able to tell the world that a reduction in atmospheric carbon dioxide levels of between a third and a half would cool the planet by about 8 degrees Fahrenheit—enough to cover most of northern Europe, and certainly every scrap of his native Sweden, in ice.

Arrhenius had no idea if his calculations reflected what had actually happened in the ice ages. There could have been other explanations, such as a weakening sun. It was another eighty years before researchers analyzing ancient air trapped in the ice sheets of Greenland and Antarctica found that ice-age air contained just the concentrations of carbon dioxide that Arrhenius had predicted. But as he reached the end of his calculations, Arrhenius also became intrigued by the potential of rising concentrations of greenhouse gases, and how they might trigger a worldwide warming. He had no expectation that this was going to happen, but it was the obvious counterpart to his first calculation. And he concluded that a doubling of atmospheric carbon dioxide would raise world temperatures by an average of about 10°F.

How did he do these calculations? Modern climate modelers, equipped with some of the biggest supercomputers, are aghast at the labor involved. But in essence, his methods were remarkably close to theirs. Arrhenius started with some basic formulae concerning the ability of greenhouse gases to trap heat in the atmosphere. These were off the shelf from Tyndall and Fourier. That was the easy bit. The hard part was deciding how much of the solar radiation Earth's surface absorbed, and how that proportion would alter as Earth cooled or warmed owing to changes in carbon dioxide concentrations.

Arrhenius had to calculate many things. The absorption capacity of different surfaces across the globe varies, from 20 percent or less for ice to more than 80 percent for dark ocean. The capacities for dark forest and light desert, grasslands, lakes, and so on lie between these two extremes. So, armed with an atlas, Arrhenius divided the surface of the planet into small squares, assessed the capacity of each segment to absorb and reflect solar radiation, and determined how factors like melting ice or freezing ocean would alter things as greenhouse gas concentrations rose or fell. Eventually he produced a series of temperature predictions for different latitudes and seasons determined by atmospheric concentrations of carbon dioxide.

It was a remarkable achievement. In the process he had virtually invented the theory of global warming, and with it the principles of modern climate modeling. Not only that: his calculation that a doubling of carbon dioxide levels would cause a warming of about 10°F almost exactly mirrors the Intergovernmental Panel on Climate Change's most recent assessment, which puts 10.4°F at the top of its likely warming range for a doubling of carbon dioxide levels.

Arrhenius presented his preliminary findings, "On the Influence of Carbonic Acid in the Air upon the Temperature of the Ground," to the Stockholm Physical Society in December 1895 and, after further refinements, published them in the *London, Edinburgh and Dublin Philosophical Magazine and Journal of Science.* There he offered more predictions that are reproduced by modern computer models. High latitudes would experience greater warming than the tropics, he said. Warming would also be more marked at night than during the day, in winter than in summer, and over land than over sea.

But he had cracked an issue that seemed to interest no one else. The world forgot all about it. Luckily for Arrhenius, this labor was but a sideshow in his career. A few years after completing it, he found fame as the winner of the 1903 Nobel Prize for Chemistry, for work on the electrical conductivity of salt solutions. Soon, too, he had a new wife and a child, and other interests—he dabbled in everything from immunology to electrical engineering. He was an early investigator of the northern lights and a popular proponent of the idea that the seeds of life could travel through space.

But after the First World War, his mood changed. The optimism of his generation, which believed that science and technology could solve every problem, crumbled in the face of a war that killed so many of its sons. He railed against the wastefulness of modern society. "Concern about our raw materials casts a dark shadow over mankind," he wrote, in an early outburst of twentieth century environmental concern. "Our descendants surely will censure us for having squandered their just birthright." His great fear was that oil supplies would dry up, and he predicted that the United States might pump its last barrel as early as 1935. He advocated energy efficiency and proposed the development of renewable energy, such as wind and solar power. He sat on a government commission that made Sweden one of the first countries to develop hydroelectric power.

Many Swedes today see Arrhenius as an environmental pioneer and praise his efforts to promote new forms of energy. He would have been bemused by this appreciation. For one thing, he never made the connection between his work on the greenhouse effect and his later nightmares about disappearing fossil fuels. He knew from early on that burning coal and oil generated greenhouse gases that would build up in the air. But he rather liked the idea, writing in 1908: "We may hope to enjoy ages with more equable and better climates, especially as regards the colder regions of the Earth, ages when the Earth will bring forth much more abundant crops for the benefit of a rapidly propagating mankind." But he had concluded with some sadness that it would probably take a millennium to cause a significant warming. And when he later began to perceive the scale of industrial exploitation of fossil fuels, his fear was solely that the resources would run out.

For half a century after Arrhenius's calculations, the prevailing view continued to be that man-made emissions of carbon dioxide were unlikely to have a measurable effect on the climate anytime soon. Nature would easily absorb any excess. From time to time, scientists did measure carbon dioxide in the air, but local variability was too great to identify any clear trends in concentrations of the gas.

The only man to take the prospect of greenhouse warming seriously was a British military engineer and amateur meteorologist, Guy Callendar. In a lecture at the Royal Meteorological Society in 1938, he said that the few existing measurements of carbon dioxide levels in the atmosphere suggested a 6 percent increase since 1900, that this must be due to fossil fuel burning, and that the implication was that warming was "actually occurring at the present time." Like Arrhenius, Callendar thought this on balance rather a good thing. And like Arrhenius, he saw his findings pretty much ignored.

The next person to make a serious effort was Charles David Keeling, a young student at the Scripps Institution of Oceanography, in La Jolla, California. He began monitoring carbon dioxide levels in the mid-1950s, first in the bear-infested hills of the state's Yosemite National Park, where he liked to go hiking, and later, in the hope of getting better data, in the clean air 14,000 feet up on top of Mauna Loa, a volcano in Hawaii. Keeling took measurements every four hours on Mauna Loa, in the first attempt ever to monitor carbon dioxide levels in one place continuously. He was so serious about his measurements that he missed the birth of his first child in order to avoid any gaps in his logbook.

The results created a sensation. Keeling quickly established that in such a remote spot as Mauna Loa, above weather systems and away from pollution, he could identify a background carbon dioxide level of 315 parts per million (ppm). The seasonal cycling of carbon dioxide caused an annual fluctuation around this average between summer and winter. Plants and other organisms that grow through photosynthesis consume carbon dioxide from the air, especially in spring. But during autumn and winter, photosynthesis largely stops, and the photosynthesizers are eaten by soil bacteria, fungi, and animals. They exhale carbon dioxide, pushing atmospheric levels back up again. Because most of the vegetation on the planet is in the Northern Hemisphere, the atmosphere loses carbon dioxide in the northern summer and gains it again in the winter. Earth, in effect, breathes in and out once a year.

But Keeling's most dramatic discovery was that this annual cycle was superimposed on a gradual year-to-year rise in atmospheric carbon dioxide levels—a trend that has become known as Keeling's curve. The background concentration of 315 ppm that Keeling found on Mauna Loa in 1958 has risen steadily, to 320 ppm by 1965, 331 ppm by 1975, and 380 ppm today.

The implications of Keeling's curve were profound. "By early 1962," he later wrote, "it was possible to deduce that approximately half of the CO_2 from fossil fuel burning was accumulating in the air," with the rest absorbed by nature. By the late 1960s he had noticed that the annual cycling of carbon dioxide was growing more intense. And the spring downturn in atmospheric levels was beginning earlier in the year—strong evidence that the slow annual increase in average levels was raising temperatures and creating an earlier spring.

Keeling personally supervised the meticulous measurements on Mauna Loa until his death, in 2005. In his final year, this generally mild man picked up the public megaphone one last time to warn that, for the first time in almost half a century, his instruments had recorded two successive years, 2002 and 2003, in which background carbon dioxide levels had risen by more than 2 ppm. He warned that this might be because of a weakening of the planet's natural ability to capture and store carbon in the rainforests, soils, and oceans—nature's "carbon sinks." He feared that nature, which had been absorbing half the carbon dioxide emitted by human activity, might be starting to give it back—something that, in his typically understated way, he suggested "might give cause for concern."

On his death, Keeling's bosses at Scripps were kind enough to call the Keeling curve "the single most important environmental data set taken in the 20th century." Nobody disagreed. One writer called him the man who "measured the breathing of the world."

Thanks to Keeling's curve, the ideas of Arrhenius and Callendar were rescued from the dustbin of scientific history. It seemed he was right that people could tamper with the planetary thermostat. Climatologists, many of whom had predicted in the 1960s that natural cycles were on the verge of plunging the world into a new ice age, began instead to warn of imminent man-made global warming. As late as the early 1970s, U.S. government officials had been asking their scientists how to stop the Arctic sea ice from becoming so thick that nuclear submarines could not break through. But by the end of the decade, President Jimmy Carter's Global 2000 Report on the environment had identified global warming as an urgent new issue, and the National Academy of Sciences had begun the first modern study of the problem.

A vast amount of research has been conducted since. For the past decade and a half, the IPCC has produced regular thousand-page updates just to review the field and pronounce on the scientific consensus. But in some ways, mainstream thinking on how climate will alter as carbon dioxide levels rise has not advanced much in the century since Arrhenius. Thanks to Keeling, we know that those levels are rising; but little else has changed.

Only in the past five years, as researchers have learned more about the way our planet works, have some come to the conclusion that changes probably won't be as smooth or as gradual as those imagined by Arrhenius—or as the scenarios of gradual change drawn up by the IPCC still suggest. We are in all probability already embarked on a roller-coaster ride of lurching and sometimes brutal change. What that ride might feel like is the central theme of this book.

2

TURNING UP THE HEAT

A skeptic's guide to climate change

Ever since the rise of concern about climate change during the 1980s, the scientists involved have been dogged by a small band of hostile critics. Every time they believe they have seen them off, the skeptics come right back. And in some quarters, their voices remain influential. One leading British newspaper in 2004 called climate change a "global fraud" based on "left-wing, anti-American, anti-West ideology." And the best-selling author Michael Crichton, in his much-publicized novel *State of Fear,* portrayed global warming as an evil plot perpetrated by environmental extremists.

Many climate scientists dismiss the skeptics with a wave of the hand and return to their computer models. Most skeptics, they note, fall into one of three categories: political scientists, journalists, and economists with little knowledge of climate science; retired experts who are aggrieved to find their old teachings disturbed; and salaried scientists with overbearing bosses to serve, such as oil companies or the governments in hock to them. If the skeptics are to be believed, the evidence for global warming and even the basic physics of the greenhouse effect are full of holes. The apparent scientific consensus exists only, they say, because it is enforced by a scientific establishment riding the gravy train, aided and abetted by politicians keen to play the politics of fear. Much of this may sound hysterical. But could the skeptics be on to something?

First, the basic physics. As we have seen, much of this goes back almost two centuries. Fourier and Tyndall both knew that the atmosphere stays warm because a certain amount of the short-wave radiation reaching Earth from the sun is absorbed by the planet's surface and radiated at longer infrared wavelengths. Like any radiator, this warms the surrounding air. They knew, too, that this heat is trapped by gases—such as water vapor, carbon dioxide, and methane—that have a "greenhouse effect," without which the planet would be frozen, like Mars. But you can have too much of a good thing. Our other planetary neighbor, Venus, has an atmosphere choked with greenhouse gases and is broiling at around 840°F as a result. And that is a worry. For, thanks to Keeling's curve, there can be no doubt now that human activity on planet Earth is raising carbon dioxide in the atmosphere to roughly a third above pre-industrial levels.

The effect this has on the planet's radiation balance is now measurable. In 2001, Helen Brindley, an atmospheric physicist at Imperial College London, examined satellite data over almost three decades to plot changes in the amount of infrared radiation escaping from the atmosphere into space. Because what does not escape must remain, heating Earth, this is effectively a measure of how much heat is being trapped by greenhouse gases—the greenhouse effect. In the part of the infrared spectrum trapped by carbon dioxide—wavelengths between 13 and 19 micrometers—she found that less and less radiation is escaping. The results for the other greenhouse gases were similar.

These findings alone should be enough to establish for even the most diehard skeptic that man-made greenhouse gas emissions are making the atmosphere warmer. Climate models developed by the U.S. government's space agency, NASA, estimate that Earth is now absorbing nearly one watt more than it releases per 10.8 square feet of its surface. This is a significant amount. You could run a 60-watt light bulb off the excess energy supplied to the area of the planet that a modest house occupies.

More contentious is whether we can actually feel the heat. Direct planet-wide temperature records go back 150 years. They suggest that nineteen of the twenty warmest years have occurred since 1980, and that the five warmest years have all been since 1998. Could the thermometers be misleading us? That has to be a possibility. The records, after all, are not a formal planetary monitoring system; they are just a collection of all the data that happen to be available.

Two important criticisms are made. One is that satellite sensors and instruments carried into the atmosphere aboard weather balloons do not back up the surface thermometers. The instrument data suggest that if air close to the surface is warming, that warming is not spreading through the bottom 6 miles of the atmosphere, known as the troposphere, in the way that climate scientists predict. If true, this is very worrying, says Steve Sherwood, a meteorologist at Yale University and author of a study of the problem: "It would spell trouble for our whole understanding of the atmosphere."

Not surprisingly, skeptics have given great play to the suggestion that satellites "prove" the surface thermometers to be at fault. Not so fast, says Sherwood. The satellite data are untrustworthy, because they measure the temperature in the air column beneath a satellite and cannot easily distinguish between the troposphere, which is expected to be warming, and the stratosphere, which should be cooling as less heat escapes the lower atmosphere. Further, satellites do not provide direct measurements in the way that thermometers do. Temperatures have to be interpreted from other data, which creates errors. The scientists running the instruments accept that the results "drift." Every week, says Sherwood, they recalibrate their satellite measurements according to data from weather balloons. In effect, therefore, the long-term average data from satellites are creatures of the balloon data.

So how good is the balloon data? Here Sherwood found a surprisingly obvious flaw—obvious, at any rate, to anyone who has left an ordinary thermometer out in the sun. The sun's ultraviolet rays shining on the bulb force the temperature reading continuously upward so that it no longer measures the air temperature. The true air temperature can be captured only in the shade, unmolested by the sun's direct rays. Thermometers on weather balloons, it turns out, are no different. They are "basically cheap thermometers easily read by an electric circuit," says Sherwood. They, too, show spurious readings when in the sun.

Meteorologists have recently fixed the problem by shielding the thermometers attached to weather balloons inside a white plastic housing. But this was rarely done thirty years ago. Sherwood concludes that "back in the 1960s and 1970s especially, the sun shining on the instruments was making readings too high." And that, he says, is the most likely explanation for why balloon measurements do not reveal a warming trend.

Two further observations back up this interpretation. First, spurious readings should not be a problem when the sun goes down, so 1960s and 1970s readings at night should be reliable. And sure enough, nighttime balloon data over the past thirty years show a warming trend. Second, the data from both balloons and satellites show a strong cooling in the stratosphere—which is likely only if more heat is truly being trapped beneath it, in the troposphere.

Another serious criticism of the surface-temperature trends is that measurements by surface thermometers have been biased by the growth of cities. The concrete and tarmac of cities retain more heat than rural areas, especially at night. The argument is that over the decades, more and more temperature-measuring sites have become urban, so the temperature trends reflect the urbanization of thermometers rather than real warming. The "urban heat island," as researchers call it, is undoubtedly real. Cities do hang on to heat. But is it skewing the global data?

This seems unlikely. The largest areas of warming have been recorded over the oceans, and the greatest magnitude of warming is mainly in polar regions, distant from big centers of population. The skeptics should finally have been silenced by a neat piece of research in 2004 by David Parker, of the Hadley Centre for Climate Prediction, part of Britain's Met Office in Exeter. He figured that the urban heat island effect should be most intense when there is no wind to disperse the urban heat. So he divided the historical temperature data into two sets: one of temperatures taken in calm weather, and the other of temperatures taken in windy weather. He found no difference. So, while nobody denies that the urban heat island effect exists, it is not sufficient to upset the reliability of global trends in thermometer readings.

There are other disputes, which we might call "second order," because they are about circumstantial evidence of climate change. Is it true, for instance, that temperatures at the end of the twentieth century were really hotter than at any other time in the past millennium? That is the claim made by U.S. researcher Michael Mann. He produced a controversial graph dubbed the "hockey stick," which used data from tree rings and other "proxy" sources to show that the millennium comprised 950 years of stable temperatures and a sudden upturn at the end. The arguments, which we will look at in more detail later, continue as to whether Mann's data are correct. And in the end, we may simply never know enough about past temperatures to be sure. But however the dispute goes, it doesn't change the basic science of the greenhouse effect. And in any event, it should be no part of the case for future climate change that past climate did not vary. It rather obviously did. As this book will argue, there is no comfort in past variability. Quite the contrary.

Similarly, there is room for uncertainty about the cause of the rise in temperature over the past 150 years, which is, depending on how you draw your average for recent years, put at a global average of between 1.1 and 1.4°F. The warming itself is real enough, but that doesn't necessarily mean that humans are to blame. It could be natural.

One argument is that more radiation reaching us from the sun can account for most of the warming of the past 150 years. This case was made best by the Danish scientists Knud Lassen and Eigil Friis-Christensen in 1991. They found a correlation between sunspot activity, which historically reflects the energy output of the sun, and temperature changes on Earth from 1850 onward. Time-based statistical correlations are notoriously tricky, because they can happen by chance; but the Danes' correlation looked convincing, and prominent skeptics took up the case. However, newer data have convinced Lassen that solar activity cannot explain more recent climate change. Declining sunspot activity since 1980 should have reduced temperatures on Earth. Instead, they have been rising faster than ever.

Overall, this particular dispute has been good for science, and the skeptics can claim a tie. Climate scientists who once put all global warming since 1850 down to the greenhouse effect now concede that up to 40 percent was probably due to the sun. Solar changes may have been the main cause of the substantial global warming in the first half of the twentieth century, for instance. But there is no way the sun's activity can explain the dramatic warming since 1970.

Both sides play one last trick. Web sites run by skeptics regularly publish temperature graphs from particular places that show no warming, suggesting that the whole idea of global warming is a myth. But climate scientists are almost as guilty when they indiscriminately attribute every local warming to global trends, whereas well-understood local climate cycles may be the more likely cause. The case for setting up local climate "watchtowers" in parts of the planet known to be sensitive to climate change, such as the Arctic, remains strong. But they will never provide unambiguous proof of global change, because global warming has not canceled out natural variations in local climate systems. What is so remarkable about recent trends is not local events but the global reach of warming. Virtually no region of the planet is spared. This is in contrast to natural oscillations that mostly just redistribute heat. The greenhouse effect is putting more energy into the entire climate system. Occasionally that causes cooling and other weird weather, but mostly it causes strong warming.

To summarize the current state of affairs: the global trends are real. No known natural effect can explain the global warming seen over the past thirty years. In fact, natural changes like solar cycles would have caused a marginal global cooling. Only some very convoluted logic can avoid the conclusion that the human hand is evident in climate change. Indeed, to think anything else would be to flout one of the central tenets of science. The fourteenth-century English philosopher William Ockham coined the principle of Ockham's razor when he argued that, if the evidence supported them, the simplest and least convoluted explanations for events were the best. Changes in greenhouse gases are the simple, least convoluted explanation for climate change. And those changes are predominantly man-made.

This is not the end of the story, however. While we can be fairly certain that more greenhouse gases in the air will push the atmosphere to further warming, big uncertainties remain about how the planet will respond. An assessment of the sensitivity of global temperatures to outside forcing —whether to changes in sunlight or the addition of greenhouse gases— mostly revolves around disentangling the main feedbacks: the things changed by an altered climate that influence the climate in turn. Positive feedbacks reinforce and amplify the change, and run the risk of producing a runaway change—the climatic equivalent of a squawk on a sound system. Negative feedbacks work in the other direction, moderating or even neutralizing change.

The current climate models concur with Arrhenius that the planet will amplify the warming. But skeptics believe that nature has strong stabilizing forces that will act as negative feedbacks and head off climate change. They don't by any means agree on how this will work. Some say a warmer world will be a cloudier world, providing us with more shade from the sun. Others, like the respected Massachusetts Institute of Technology meteorologist Richard Lindzen, have argued that the higher reaches of the troposphere might actually become drier, reducing the greenhouse effect of water vapor. Many of these arguments reflect legitimate uncertainty among climate scientists, though some of the negative feedbacks proposed by the skeptics, such as cloud processes, could equally turn into major positive feedbacks and make the IPCC projections too small.

Where does this leave us? Actually, with a surprising degree of scientific consensus about the basic science of global warming. When the science historian Naomi Oreskes, of the University of California in San Diego, reviewed almost a thousand peer-reviewed papers on climate change published between 1993 and 2003, she found the mainstream consensus to be real and near universal. "Politicians, economists, journalists and others may have the impression of confusion, disagreement or discord among climate scientists, but that impression is incorrect," she concluded. The disagreements were mainly about detail. The consensus, stretching from Tyndall through Arrhenius to the IPCC, lived on.

For hard-line skeptics, of course, any scientific consensus must, by definition, be wrong. As far as they are concerned, the thousands of scientists behind the IPCC models have either been seduced by their own doom-laden narrative or are engaged in a gigantic conspiracy. For them, the greater the consensus, the worse the conspiracy. The maverick climatologist Pat Michaels, of the University of Virginia in Charlottesville, says we are faced with what the philosopher of science Thomas Kuhn called a "paradigm problem." Michaels, who is also the state meteorologist for Virginia, one of the United States' largest coal producers, and a consultant to numerous fossil fuel companies, says: "Most scientists spend their lives working to shore up the reigning world view—the dominant paradigm— and those who disagree are always much fewer in number." The drive to conformity, he says, is accentuated by peer review, which ensures that only papers in support of the paradigm appear in the research literature, and by public funding of research into the prevailing "paradigm of doom."

Even if you accept this cynical view of how science is done, it doesn't mean that the orthodoxy is always wrong. The fact that scientists universally agree that the world is round does not make it flat. Many of the same claims that are now made against the global warming "paradigm" were once made about the "AIDS industry" by people who disputed that HIV caused AIDS. Some governments took their side for a long time, and their citizens are now living with the consequences. Where are those skeptics now? Some of them can be heard making the case against climate change.

But all that said, I do think the skeptics are important to the arguments about climate science. The desire for consensus is always likely to lead the mainstream scientific community to don blinkers. This has not only blotted out the arguments of skeptics but also sidelined results from the handful of "rogue" climate models that keep turning up tipping points that could tumble the world into much worse shape than what is currently predicted by the mainstream. One scientist told me in the corridors of a conference in early 2005: "By ignoring these outliers, IPCC has failed for ten years to investigate the possible effects of more extreme climate change."

So, despite their sometimes cynical motives, the skeptics have served a purpose in picking away at the IPCC orthodoxy. As in politics, every good government needs a good opposition. And though their arguments have often been opportunistic and personal, the skeptics have spotted the stifling impact of consensus-building. They are, if nothing else, helping to keep the good guys honest. The pity is that they have not done a better job, by engaging in more real science and less empty rhetoric. And in their enthusiasm to debunk climate change, they have failed to grasp one alarming possibility: that the IPCC could be underestimating, not overestimating, the threat that the world faces.

3
THE YEAR
How the wild weather of 1998 broke all records

Lidia Rosa Paz was at a loss. She caught my arm and pointed despairingly into the raging river. Out there, about 50 yards into the water, was the spot where, until days before, she had lived. On the night of October 28, 1998, her shantytown of Pedro Dias, in the town of Choluteca, in Honduras, had been washed away, taking more than a hundred people to their deaths. Lidia had survived, but every one of her possessions was gone. "What will I do now?" she asked. I didn't have an answer.

Hers was one story from a night when floods and landslides ripped apart the small Central American country's geography, leaving more than 10,000 Hondurans dead and 2 million homeless. It was the night that Hurricane Mitch, the most vicious hurricane to hit the Americas in 200 years, came calling, and dumped a year's rain in just a few hours. Choluteca is in southern Honduras, on the Pacific coast, far from the normal track of Caribbean hurricanes. When the radio issued storm warnings that night, neither Lidia nor any of her neighbors took much notice. "Hurricanes never come here," she told me. Or at least they never had.

I was in Honduras a couple of weeks after the hurricane had struck. The devastation was appalling. Huge floods had rushed down rivers and into the capital, Tegucigalpa, in the mountainous heart of the country, ripping away whole communities. A thousand people lost their lives beneath a single slide that landed on the suburb of Miramesi. Another stopped just short of the American embassy in the capital. Rivers changed their paths right across the country, obliterating towns. And flash floods on steep hillsides buried whole communities under mud. Sixty percent of the country's bridges were destroyed, along with a quarter of its schools and half its agricultural productivity, including nearly all its banana plantations. The first visitors to the southern town of Mordica reported, "All you can see is the top of the church." Ministers said the country's economic development had been put back twenty years.

For tens of millions of people across the world, the violence of Mitch is an omen. Many climatologists believe that Mitch, a ferocious hurricane made worse by the warm seas that allowed it to absorb huge amounts of water from the ocean, was a product of global warming—and a sign of things to come for the hundreds of millions of inhabitants of flood-prone river valleys and coastal plains across the world; for those living on deforested hillsides prone to landslips; and for many millions more who do not yet know that they are vulnerable in a new era of hyperweather. People like Lidia before Mitch hit.

Those who do not believe that global warming is a real and dangerous threat should visit places like Choluteca and talk to people like Lidia. It may not convince them that climate change is making superhurricanes and megafloods. But it will show them the forces of nature untamed and the human havoc caused when weather breaks its normal shackles. For hundreds of millions of people, these issues are no longer a matter for computer modeling or debate in the corridors of Congress or future forecasts. They are about real lives and deaths. The question is not: Can we prove that events like Mitch are caused by climate change? It is: Can we afford to take the chance that they are?

The year 1998 was the warmest of the twentieth century, perhaps of the millennium. It was also a year of exceptionally wild weather, and few doubt that the two were connected. That year, besides the storms, the rainforests got no rain. Forest fires of unprecedented ferocity ripped through the tinder-dry jungles of Borneo and Brazil, Peru and Tanzania, Florida and Sardinia. New Guinea had the worst drought in a century; thousands starved to death. East Africa saw the worst floods in half a century—during the dry season. Uganda was cut off for several days, and much of the desert north of the region flooded. Mongol tribesmen froze to death as Tibet had its worst snows in fifty years. Mudslides washed houses off the cliffs of the desert state of California. In Peru, a million were made homeless by floods along a coastline that often has no rain for years at a time. The water level in the Panama Canal was so low that large ships couldn't make it through. Ice storms disabled power lines throughout New England and

Quebec, leaving thousands without power or electric light for weeks. The coffee crop failed in Indonesia, cotton died in Uganda, and fish catches collapsed in the Pacific off Peru. Unprecedented warm seas caused billions of the tiny algae that give coral their color to quit reefs across the Indian and Pacific Oceans, leaving behind the pale skeletons of dead coral.

All a coincidence? Not according to the IPCC. Some of the damage was caused by an intense outbreak of a natural climate cycle in the Pacific known as El Niño. Every few years, this causes a reversal of winds and ocean currents across the equatorial Pacific, for a few months taking rains to drought regions and droughts to normally wet areas. But as we shall see in Chapter 30, there is growing evidence that El Niños are becoming stronger and more frequent under the influence of global warming. This is probably part of a pattern identified by the IPCC, in which, all around the world, the weather is becoming more extreme and more unpredictable as the world warms. And 1998, the warmest year yet, was the epitome of the trend.

The heat is intensifying the hydrological cycle. Globally, average annual rainfall increased by up to 10 percent during the twentieth century, because warming has increased evaporation. Locally, the trends are even stronger. The floods that inundated Mozambique in 2000 occurred because maximum daily rainfall there had risen by 50 percent. In the eastern U.S., the proportion of rain falling in heavy downpours has increased by a quarter. In Britain, winter rain falls in intense downpours twice as often as it did in the 1960s. There are similar patterns in Australia, South Africa, Japan, and Scandinavia. Even the Asian monsoon has become more intense but less predictable. At the same time, dry areas in continental interiors have become drier, causing deserts to spread. The year 1998 was the first in a run of years of intense drought that stretched from the American West through the Mediterranean to Central Asia.

At the time of this writing, no other year has been as hot as 1998—and no other year so climatically violent. Unless, that is, you were caught in one of the record number of tropical storms in the North Atlantic in 2005. But if you want to know what the first stage of climate change is shaping up to be like, look no further than 1998.

4

THE ANTHROPOCENE

A new name for a new geological era

Welcome to the Anthropocene. It's a new geological era, so take a good look around. A single species is in charge of the planet, altering its features almost at will. And what more natural than to name this new era after that top-of-the-heap anthropoid, ourselves? The term was coined in 2000 by the Nobel Prize–winning Dutch atmospheric scientist Paul Crutzen to describe the past two centuries of our planet's evolution. "I was at a conference where someone said something about the Holocene, the long period of relatively stable climate since the end of the last ice age," he told me later. "I suddenly thought that this was wrong. The world has changed too much. So I said: 'No, we are in the Anthropocene.' I just made up the word on the spur of the moment. Everyone was shocked. But it seems to have stuck."

The word is catching on among a new breed of scientists who study Earth systems—how our planet functions. Not just climate systems, but also related features, such as the carbon cycle on land and at sea, the stratosphere and its ozone layer, ocean circulation, and the ice of the cryosphere. And those scientists are coming to believe that some of these systems are close to breakdown, because of human interference. If that is true, then the gradual global warming predicted by most climate models for the next centuries will be the least of our worries.

The big new discovery is that planet Earth does not generally engage in gradual change. It is far cruder and nastier, says Will Steffen, an Australian expert on climate and carbon cycles who from 1998 to 2004 was director of the International Geosphere Biosphere Programme, a research agency dedicated to investigating Earth systems. A mild-mannered man not given to hyperbole, Steffen nonetheless takes a hard-nosed approach to climate change. "Abrupt change seems to be the norm, not the exception," he says. We have been lured into a false sense of security by the relatively quiet climatic era during which our modern complex civilizations have grown and flourished. It may also have left us unexpectedly vulnerable as we stumble into a new era of abrupt change.

We have also been blind, he says, to the extent of the damage we are doing to our planetary home. We often see our impact as limited to individual parts of the system: to trashed rainforests, polluted oceans, and even raised air temperatures. We rarely notice that by doing all these things at once, we are undermining the basic planetary systems. Something, Steffen says, is going to give: "The planet may have an Achilles heel. And if it does, we badly need to know about it." Without that knowledge and the will to act, he says, the Anthropocene may well end in tears.

A report from the U.S. National Academy of Sciences in 2002, under the chairmanship of Richard Alley, of Penn State University—a glaciologist with the slightly manic appearance of an ex-hippie, who has become a regular on Capitol Hill for his ability to talk climate science in plain language—sounded a similar warning. "Recent scientific evidence shows that major and widespread climate changes have occurred with startling speed," the report began. "The new paradigm of an abruptly changing climate system has been well established by research over the last decade, but this new thinking is little known and scarcely appreciated in the wider community of natural and social scientists and policymakers." Or, Alley might have added, among the citizens of this threatened planet.

We have already had one lucky break. It happened twenty years ago, when a hole suddenly opened in the ozone layer over Antarctica, stripping away the continent's protective shield against ultraviolet radiation. We were lucky that it happened over Antarctica, and lucky that we spotted it before it spread too far.

Many of the scientists who worked to unravel the cause of the ozone hole—including Crutzen, who won his Nobel Prize in this endeavor—are among the most vehement in issuing the new warnings. They know how close we came to disaster. Glaciologists like Alley are another group who take the perils of the Anthropocene most seriously. In the past decade, they have analyzed ice cores from both Greenland and Antarctica to map the patterns of past natural climate change. The results have been chilling.

It has emerged, for instance, that around 12,000 years ago, as the last ice age waned and ice sheets were in full retreat across Europe and North America, the warming abruptly went into reverse. For a thousand years the world returned to the depths of the ice age, only to emerge again with such speed that, as Alley puts it, "roughly half of the entire warming between the ice ages and the postglacial world took place in only a decade." The world warmed by at least 9 degrees—the IPCC's prediction for the next century or so—within ten years. This beggars belief. But Alley and his co-researchers are adamant that the ice cores show this happened.

Similar switchback temperature changes occurred regularly through the last glaciation, and there were a number of other "flickers" as the planet staggered toward a new postglacial world. Stone Age man, with only the most rudimentary protection from a climatic switchback, must have found that tough. Heaven knows how modern human society would respond to such a change, whereby London would have a North African climate, Mexican temperatures would be visited on New England, and India's billion-plus population would be deprived of the monsoon rains that feed them.

The exact cause of the rise and fall of the ice ages still excites disputes. But it seems that the 100,000-year cycles of ice ages and interglacials that have persisted for around a million years have coincided with a minor wobble in Earth's orbit. Its effect on the solar radiation reaching the planet is minute, and it happens only gradually. But somehow Earth's systems amplify its impact, turning a minor cooling into an abrupt freeze or an equally minor warming into a sudden defrost. The amplification certainly involves greenhouse gases, as Arrhenius long ago surmised. The extraordinary way in which temperatures and carbon dioxide levels have moved in lockstep permits no other interpretation. It also probably involves changes to ocean currents and the temperature feedbacks from growing and melting ice.

We will return to this conundrum later. What matters here is that a minor change in the planet's heating—much less, indeed, than we are currently inflicting through greenhouse gases—could cause such massive changes worldwide. The planet seems primed to leap into and out of glaciations and, perhaps, other states too.

Some see this hair trigger as rather precisely organized. Will Steffen says that for a couple of million years, Earth's climate seems to have had just two "stable states": glacial and interglacial. There was no smooth transition between them. The planet simply jumped, at a signal from the orbital wobble, from the glacial to the interglacial state, and made the jump back again with a little, but not much, more decorum. "The planet jumps straight into the frying pan and makes a bumpy and erratic slide into the freezer," Steffen says. The glacial state seems to have been anchored at carbon dioxide levels of around 190 ppm, while the interglacial state, which the modern world occupied until the Industrial Revolution, was anchored at about 280 ppm. The rapid flip between the two states must have involved a reallocation of about 220 billion tons of carbon between the oceans, land, and the atmosphere. Carbon was buried in the oceans during the glaciations and reappeared afterward. Nobody knows quite how or why. But the operation of the hair-trigger jump to a much warmer state raises critical questions for the Anthropocene.

In the past two centuries, humanity has injected about another 220 billion tons of carbon into the atmosphere, pushing carbon dioxide levels up by a third, from the stable interglacial level of 280 ppm to the present 380 ppm. The figure continues to rise by about 20 ppm a decade. So the big question is how Earth will respond. Conventional thinking among climate scientists from Arrhenius on predicts that rising emissions of carbon dioxide will produce a steady rise in atmospheric concentrations and an equally steady rise in temperatures. That's still the IPCC story. But Steffen takes a different view: "If the ice age seemed to gravitate between two steady states, maybe in future we will gravitate to a third steady state." Nature might, he concedes, fulfill the expectations of climate skeptics and push back down toward 280 ppm; but if it was going to do that, we would already see evidence of it. And we don't.

Other scientists, including Alley, are not convinced by Steffen's sense of order in the system. Sitting in his departmental office, Alley likens the climate system to "a drunk—generally quiet when left alone, but unpredictable when roused." When he is writing scientific papers or committee reports, his language is not so vivid. He talks of a "chaotic system" vulnerable to "forcings" from changes in solar radiation or greenhouse gases. "Abrupt climate change always could occur," he says. But "the existence of forcings greatly increases the number of possible mechanisms [for] abrupt change"; and "the more rapid the forcings, the more likely it is that the resulting change will be abrupt on the timescale of human economies or global ecosystems." Drunks, in other words, may be unpredictable, but if you shout at them louder or push them harder, they will react more vehemently. Right now, moreover, we are offering our drunk one more for the road.

The past 10,000 years, since the end of the last ice age, have not been without climate change. The Asian monsoon has switched on and off; deserts have come and gone; Europe and North America have flipped from medieval warm period to little ice age. None of these events has been as dramatic as the waxing and waning of the ice ages themselves. But most were equally abrupt, and civilizations have come and gone in their wake. Even so, human society in general has prospered, learning to plant crops, domesticate animals, tame rivers, create cities, develop science, and ultimately industrialize the planet.

But in the Anthropocene, the rules of the game have changed. Alley and Steffen agree that humanity is today pushing planetary life-support systems toward their limits. The stakes are higher, because what is happening is global. "Before, if we screwed up, we could move on," says Steffen. "But now we don't have an exit option. We don't have another planet."

5

THE WATCHTOWER

Keeping climate vigil on an Arctic island

A chill wind was blowing off the glacier. Small blue chunks of ice occasionally split from its face and floated down the fjord toward the ocean. A strange green ribbon of light flashed across the sky above from an anonymous building on the foreshore. And on the snow behind, a polar bear wandered warily around a strange human settlement that had grown up on this remote fjord at the seventy-ninth parallel.

I had come to Ny-Alesund, an international community of scientists that, in the darkening days of autumn, numbered fewer than thirty people. The hardy band was there to man this Arctic watchtower on the northwest shores of Spitzbergen, the largest island of a cluster of Arctic islands called Svalbard, because it is reckoned to be one of the most likely places to witness firsthand any future climatic conflagration. Hollywood directors may have chosen New York as the place that would descend into climatic chaos first. But while the scientists here heartily enjoy watching their DVD of *The Day After Tomorrow,* they are convinced that Ny-Alesund is the place to be. The place where our comfy, climatically benign world might begin to end. Where nature may start to take its revenge.

Ny-Alesund is a tiny town of yellow, red, and blue houses two hours' flight from the northernmost spot on mainland Europe. It is nearer Greenland and the North Pole than Norway, which administers Svalbard under an international treaty signed in 1920. It has history. This was where great Norwegian Arctic explorers such as Roald Amundsen and Graf Zeppelin set out for the North Pole, by ship, seaplane, and even giant airship assembled here. More recently, the High Arctic was famous for its military listening posts, where the staff sat in the cold silence, waiting for the first sign of a Russian or American nuclear missile streaking over the ice to obliterate New York or Moscow or London. But today the biggest business is climate science—waiting for the world to turn. Says Jack Kohler, of the Norwegian Polar Institute, down south in Tromso: "If you want to see the world's climate system flip, you'd probably best come here to see it first."

Spitzbergen is already one of the epicenters of climate change. For a few days in July 2005, the scientists put aside their instruments, donned T-shirts and shorts, and sipped lager by the glaciers in temperatures that hit a record 68°F—just 600 miles from the North Pole. Even in late September, as the sun hovered close to the horizon and the long Arctic night beckoned, the sea was still ice-free, and tomatoes were growing in the greenhouse behind the research station kitchens. Old-timers like the British station head Nick Cox, who has visited Ny-Alesund most years since 1978, marvel at the pace of change. "It stuns me how far the glaciers have retreated and how the climate has changed," Cox says. "It used to be still and clear and cold. Now it is a lot warmer, and damper, too, because the warmer air can hold more moisture."

Photographs in the town's tiny museum show families who used to work in coal mines here in the 1930s, huddled in warm clothes down by the shore. Looming behind them are glaciers that are barely visible today, having retreated about 3 miles back up the fjord. The glaciers and ice sheets that still cover two thirds of Svalbard are some of the best-studied in the world. And visiting glaciologists leave each time with worsening news. In the summer of 2005, British glaciologists discovered that the nearby Midtre Lovenbreen glacier had lost 12 inches of height in a single week as it melted in the sun. The Kronebreen glacier may be dumping close to 200,000 acre-feet of ice into the fjord every year.

Jack Kohler is attempting a "mass balance" of the ice of Svalbard. He reckons that 20 million acre-feet melts and runs off into the ocean each year now. Another 3 million acre-feet is lost from icebergs slumping into the sea from 620 miles of ice cliffs. At most, half of this loss is being replaced with new snow. That is an annual net loss of around 11 million acre-feet— a staggering volume for a small cluster of islands, and probably second in the Arctic only to the loss from the huge ice sheet covering Greenland.

And there is more to come, Kohler says. Many of Svalbard's glaciers and ice caps are close to the freezing point and "very sensitive to quite small changes" in temperature. Boreholes drilled into the permafrost show a staggering 0.7°F warming in the past decade. A few more tenths of a degree could be catastrophic, he says.

Ny-Alesund is a cosmopolitan community, especially in summer, with Norwegians and Germans, Swedes and British, Spanish and Finns, Italians and French, Russians and Americans, Japanese and Chinese and Koreans. It is also quirky. Checking some equipment in the empty Korean labs, I found a pair of Spanish scientists hiding there. They said they couldn't afford the accommodation fees in the main compound, but couldn't bear to give up their work measuring glaciers. The Chinese had departed for the winter, but left behind a pair of two-ton granite lions to guard the entrance to their building. The week before, a shipload of Scotsmen, dressed in kilts and offering whiskey galore, showed up at the quayside for some R&R while investigating the sediments on the bottom of the fjord; and since then some Yorkshiremen had flown a remote-controlled helicopter the size of a small dog over glaciers to map them in 3D.

At Ny-Alesund there are magnetometers and riometers and spectrophotometers probing the upper atmosphere; there are weather balloons aplenty, a decompression chamber for divers, and even a big radio telescope that measures the radiation from distant quasars with such accuracy that it helps correct global positioning systems for the effects of continental drift. The scientists here measure chlorofluorocarbons (CFCs) and carbon dioxide, mercury and ozone, water vapor and radon; they fingerprint the smoke and dust brought in on the breeze to find out where they came from; they photograph the northern lights and sniff for methane from the melting tundra. On some cloudless nights, the German researcher Kai Marholdt sends that green shaft of laser light into the sky to probe the chemistry of the stratosphere. There is so much scientific equipment littering the tundra that nobody is sure what is still in use and what has been abandoned by long-since-departed researchers. There are plans for a cleanup, because passing reindeer keep getting tangled in the cables.

Meanwhile, the bears are coming. As the sea ice disappears, polar bears that live out on the ice and hunt for seals are being forced ashore. They are becoming bold. They break into the huts dotting the island, which are maintained for scientists spending a night out on the ice. They are looking for meat, but will sink their teeth into anything soft—bed mattresses and even inflatable boats have been torn to shreds. Anyone moving out of Ny-Alesund has to carry a gun.

Svalbard has long been recognized as extremely sensitive to climate variations. In the early twentieth century, during a period of modest warming in much of the Northern Hemisphere, temperatures rose here by as much as 9°F—a figure probably not exceeded anywhere on the planet. In the 1960s they fell again by almost as much, but the rise since has taken them back to the levels of the 1920s, with no end in sight. Climatologists warn against seeing warming here as an unambiguous sign of man-made climate change. But Ny-Alesund does seem uniquely sensitive to nudges on the planetary thermostat. It is a place where climate feedbacks like melting sea ice and changes in winds and ocean currents work with special force. And who knows what the future will hold? Only about a hundred miles out to sea, Wadhams's last chimney may be living out its final days.

Svalbard is a place to watch like a hawk, and not just for changing climate. The ozone layer is on a hair trigger here, too. Many researchers expect a giant ozone hole to form over the Arctic one day soon, just as it did in the Antarctic twenty-five years ago. And so, on the roof of the Norwegian Polar Institute, the largest research station in Ny-Alesund, pride of place goes to a gleaming steel instrument with a grand embossed nameplate announcing that you are in the presence of Dr. Dobson's Ozone Spectrophotometer No. 8—Dobson Meter No. 8, for short. The British meteorologist Gordon Dobson, one of the earliest researchers into the ozone layer, built the first of his spectrophotometers in 1931, in a wooden hut near Oxford. His eighth, built in 1935, came north to Ny-Alesund and ever since has been pointing to the sky, measuring the ultraviolet radiation pouring through the atmosphere, and thus indirectly measuring the thickness of the ozone layer.

Dobson eventually produced 150 machines. They still form the core of the world's ozone-layer monitoring network. Their work was considered routine, even dull, until one of them discovered an ozone hole over Antarctica in the early 1980s. Now Dobson Meter No. 8 and its minder, research assistant Carl Petter Niesen, are looking into the skies above Ny-Alesund for a repeat here. The most northerly and among the oldest in continual service, the instrument needs a little help these days to keep going. It has a duvet and a small heater to keep it from seizing up in the winter cold. Uniquely here, it is not connected to a computer logger. Even in the depths of winter, Niesen goes up on the roof to write down its reading with a pencil in a large logbook. Not much science happens that way anymore, but the Dobson meter, with its idiosyncratic but continuous record for more than half a century, is irreplaceable.

Dobson Meter No. 8 hasn't spotted a full-blown hole in the ozone layer yet. But as the researchers have waited, they have discovered other strange things happening to the chemistry of the atmosphere. Svalbard, it turns out, is on the flight path of acid fogs from Siberia that get trapped in thin, pancakelike layers of air close to the ice and turn the clear, still air into a yellow haze. Sometimes it rains mercury here, as industrial pollution cruises north and suddenly, within a matter of minutes, precipitates onto the snow.

Pesticides, too, have arrived in prodigious quantities, apparently from the fields of Asia. They condense in the cold air and become absorbed in vegetation. They work their way up the food chain to fish and polar bears and birds. But the very highest concentrations occur in a lake on Bear Island, in the south of the Svalbard archipelago, beneath a huge auk colony. The chemicals that have become concentrated in the Arctic air, and then concentrated again in the Arctic food web, are concentrated one more time in the urine of the auks. What at first sight might seem to be just about the least polluted place on Earth turns out to be a toxic sump.

Ny-Alesund is the most northerly permanent settlement on Earth. And the summit of Mount Zeppelin, 1,600 feet above the settlement, is the top of the top of the world—the ultimate watchtower for the world's climate. I went to the summit in the world's most northerly cable car with Carl Petter Niesen, who was taking his daily journey to tend the huge array of instruments designed to sniff every molecule of passing Arctic air. Recently, he says, carbon dioxide levels in the air on Mount Zeppelin have increased more sharply than at other monitoring stations around the world. Some days he measures levels approaching 390 ppm—fully 10 ppm above the global average. There is always some scatter in the readings. But it seems, he says, as if fast-rising emissions from power plants and cars in China and India are traveling north on the winds with the mercury and the pesticides and the acid haze. Not for the first time, he has caught a whiff of the future here at the top of the world.

II

FAULT LINES IN THE ICE

6

NINETY DEGREES NORTH

Why melting knows no bounds in the far North

"Has anybody in history ever got to 90° north, to be greeted by water and not ice?" That was the question posed by a group of scientists after returning from a cruise to the North Pole in August 2000. Sailing north from Svalbard on one of the world's most powerful icebreakers, the *Yamal*, the researchers found very little ice to break. And when they got to their polar destination, they were amazed to find not pack ice but a mile-wide expanse of clear blue water.

The story went around the world. For some, it revived the tales of ancient mariners, who said that beyond the Arctic ice there was an open ocean, and beyond that a mystical land, an Atlantis of the North. The proprietors of the *Yamal* were quick to cash in, offering summer cruises to "the land beyond the pole." But for the less romantically inclined, the story of the ice-free North Pole ignited panic about Arctic melting. By chance, the scientists on board the *Yamal* had included James McCarthy, a Harvard oceanographer on summer vacation from chairing an IPCC working group on the impacts of climate change. He didn't want to be alarmist, he said on his return. The Arctic ice sheet is made up of shifting plates, so there are bound to be gaps. But there were more and more gaps. So the unexpected discovery was "a dramatic punctuation to a more remarkable journey, in which the ice was everywhere thin and intermittent, with large areas of open water."

The whole Arctic was remarkably ice-free that summer. And that included the Holy Grail of generations of Arctic explorers, the Northwest Passage. The search for a route from the Atlantic to the Pacific and the riches of the Orient excited early explorers almost as much as El Dorado.

35

But it was a deadly pursuit. The ice swallowed up hundreds of them, most notably Sir John Franklin, whose 1845 expedition disappeared with all 128 hands. But in 2000, a Canadian ship made the journey through the Northwest Passage without touching ice. Its skipper, Ken Burton, said: "There were some bergs, but we saw nothing to cause any anxiety."

Inuit whalers the previous June told glaciologists meeting in Alaska that the ice had been disappearing for some years. "Last year it stayed over the horizon the whole summer; we had to go thirty miles just to hunt seals," said Eugene Brower, of the Barrow Whaling Captain's Association. Recently declassified data from U.S. and British military submarines had revealed that the Arctic ice in late summer was on average 40 percent thinner in the 1990s than in the 1950s. And NASA satellites, which had been photographing the ice for a quarter century, offered the most incontrovertible evidence. Their analyst-in-chief is Ted Scambos, of the National Snow and Ice Data Center, in Boulder, Colorado, a wannabe astronaut who turned to exploring the polar regions as a second best. He reports annually on how the retreat of ice is turning into a rout. In 2005, just 2 million square miles of ice were left in mid-September, the usual date of minimum ice cover. That was 20 percent less than in 1978.

The Arctic is a place without half measures. There is no mid point between water and ice. Melting and freezing are, in the jargon of the systems scientists, threshold processes. Melting takes a lot of solar energy, but once it is complete, the sun is free to warm the water left behind. And, because it is so much darker, that water is also far better at absorbing the solar energy and using it to heat the ambient air. "This makes the whole ice sheet extremely dynamic," says Seymour Laxon, a climate physicist at University College London. "The concept of a slowly dwindling ice pack in response to global warming is just not right. The process is very dynamic, and it depends entirely on temperature each summer."

"Feedbacks in the system are starting to take hold," Scambos says. The winter refreeze is less complete every year; the spring melt is starting ever earlier—seventeen days earlier than usual in 2005. "With all that dark, open water, you start to see an increase in Arctic Ocean heat storage." The Arctic "is becoming a profoundly different place." Most glaciologists agree with Scambos that the root cause of the great melt is Arctic air temperatures that have risen by about 3 to 5°F in the past thirty years—several times the global average. Global warming, it seems, is being amplified here. This is partly because the feedbacks of melting ice create extra local warming. And partly, too, because of a long warm phase in a climatic variable called the Arctic Oscillation, which brings warm winds farther north into the Arctic. The Arctic Oscillation is a natural phenomenon, but there is growing evidence that it is being accentuated by global warming, as we shall see in Chapter 37.

There is another driver for the melting, again probably connected to global warming. Warmer air above the ice is being accompanied by warmer waters beneath. Weeks before Scambos published his 2005 report, Igor Polyakov, of the International Arctic Research Center, in Fairbanks, Alaska, reported on an "immense pulse of warm water" that he had been tracking since it entered the Arctic in 1999. It had burst through the Fram Strait, a narrow "throat" of deep water between Greenland and Svalbard that connects the Greenland Sea and the Atlantic to the Arctic Ocean. And since then, it had been slowly working its way around the shallow continental shelves that encircle the Arctic Ocean. One day in February 2004, the pulse reached a buoy in the Laptev Sea north of Siberia. A thermometer strapped to the buoy recorded a jump in water temperature of half a degree within a few hours. The warm water stayed, the rise proved permanent, and the Laptev Sea rapidly became ice-free. "It was as if the planet became warmer in a single day," Polyakov told one journalist.

Pulses of warm water passing through the Fram Strait may be a regular feature of the Arctic. They were known to the Norwegian explorer and oceanographer Fridtjof Nansen, who a century ago used a specially strengthened ship called the *Fram* to float with the ice and monitor currents in the Arctic. But as the Atlantic itself becomes warmer, the pulses appear to become bigger, and their impact on the Arctic is growing. One theory is that some of the water that once disappeared down the chimneys in the Greenland Sea now comes farther north into the Arctic.

"The Arctic Ocean is in transition toward a new, warmer state," says Polyakov. And most glaciologists working in the Arctic agree. Writing in the journal of the American Geophysical Union, *Eos,* in late 2005, a group of twenty-one of them began in almost apocalyptic terms: "The Arctic system is moving to a new state that falls outside the envelope of glacial-interglacial fluctuations that prevailed during recent Earth history." Soon the Arctic would be ice-free in summer, "a state not witnessed for at least a million years," they said. "The change appears to be driven largely by global warming, and there seem to be few, if any, processes within the Arctic system that are capable of altering the trajectory towards this 'super-interglacial' state."

What would the world be like with an ice-free Arctic? Oil and mineral companies and shipping magnates long for the day when they can prospect at will, build new cities, and navigate their vessels in all seasons from Baffin Island to Svalbard and Greenland and Siberia. But it would be a world without polar bears and ice-dwelling seals, a world with no place for the Inuit way of life. And the influence of such a change would spread around the world. Without the reflective shield of ice, the whole world would warm several more degrees; ocean and air currents driven by temperature differences between the poles and the tropics would falter; on land, methane and other gases would break out of the melting permafrost, raising temperatures further; and as the ice caps on land melted, sea levels would rise so high that much of the world's population would have to move or drown. If the Arctic is especially sensitive to climate change, the whole planet is especially sensitive to changes in the Arctic.

7

ON THE SLIPPERY SLOPE

Greenland is slumping into the ocean

We are on "a slippery slope to hell." That is not the kind of language you expect to read in a learned scientific paper by one of the top climate scientists in the U.S., who is, moreover, the director of one of NASA's main science divisions, the Goddard Institute for Space Studies, in New York. Not even in a picture caption. But Jim Hansen, President George W. Bush's top in-house climate modeler, though personally modest and unassuming, calls it as he sees it.

I've followed Hansen's work for a long time. He began his career investigating the greenhouse effect on Venus, and was principal investigator for the *Pioneer* space probe to that planet in the 1970s. But he soon switched to planet Earth. He was the first person to get global warming onto the world's front pages, during the long, hot U.S. summer of 1988. Half the states in the country were on drought alert, and the mighty Mississippi had all but dried up. The Dust Bowl, it seemed to many, was returning. Hansen picked that moment to turn up at a hearing of the Senate's Energy and Natural Resources Committee in Washington and tell the sweating senators: "It is time to stop waffling so much. We should say that the evidence is pretty strong that the greenhouse effect is here." He didn't quite say that greenhouse gases were causing the drought across the country—a claim that would have been hard to substantiate. But everybody assumed he had.

Sixteen years later, Hansen was the senior U.S. government employee who, seven days before the 2004 presidential election, began a public lecture with the words "I have been told by a high government official that I should not talk about dangerous anthropogenic interference with climate, because we do not know how much humans are changing the earth's climate or how much change is dangerous. Actually, we know quite a lot." And he went on to describe what we know in some detail. Most of his fellow researchers thought that would be the end for Hansen as a government employee. But a year later this outwardly diffident man—who couldn't stop apologizing for keeping me waiting when we met in his large, paper-strewn office—was still at his post. To the astonishment of many of his colleagues. "He is saved by his science; he is just too good to be fired," said one. "Also, he is one of the good guys. He doesn't have enemies. If he needed saving, there are a lot of people who would volunteer for the job."

And now Hansen says the world, or more particularly Greenland, is on a slippery slope to hell. We had better listen.

The world's three great ice sheets—one over Greenland and the other two over Antarctica—contain vast amounts of ice. Leftovers from the last ice age, they are piles of compressed snow almost 2 miles high. Glaciologists divide the sheets into two parts. On the high ground inland, where snowfall is greatest and melting is least, they accumulate ice. But on the edges and on lower ground, where snowfall is usually less and melting is greater, they lose ice. The boundary between the two zones is known as the equilibrium line.

For many centuries these great ice sheets have been in balance, with ice loss at the edges matched by accumulation in the centers, and the equilibrium line remaining roughly stationary. Glaciologists have regarded this balance as rather secure, since such huge volumes of ice can change only very slowly. Glacially. This image of stability and longevity is reassuring. If the ice sheets all melted, or slumped into the ocean, they would make a big splash. They contain enough ice to raise sea levels worldwide by 230 feet. That would drown my house, and probably yours, too. Luckily, as glaciologists have been telling us for years, this won't happen. Not even if there is fast global warming. Large ice sheets, they say, tend to maintain their own climate, keeping the air above cold enough to prevent large-scale melting. And even if warming did take hold at the surface, it could penetrate the tightly packed ice only extremely slowly.

The scariest suggestion, made by the IPCC in 2001, was that beyond a warming of about 5°F, Greenland might gradually start to melt, with a wave of warmth moving down through the ice. Once under way, the process might be unstoppable, because as the ice sheet melted, its surface would lower and become exposed to ever-warmer air. But the melting would take place very slowly, "during the next thousand years or more." Now, that is not a nice legacy to leave to future generations, but a thousand years is forty or so generations away. So maybe it is not something to worry us today.

That used to be the scientific consensus. But Hansen is the spokesperson for a growing body of glaciologists who say that things could happen much faster. Because ice sheets, even the biggest and slowest and most stable-looking, have a secret life involving dramatic and dynamic change. And their apparent stability could one day be their undoing. The story is told best in a single picture. Hansen's "slippery slope" caption accompanied a photograph of a river of water flowing across the Greenland ice sheet and pouring down a hole. The photo has an apocalyptic feel, and in the top right-hand corner a couple of researchers look on from a distance, giving an awesome sense of scale.

What is going on here? The water is not entirely new. Small lakes have always formed on the surface of Greenland ice in the summer sun. And sometimes those lakes empty down flaws in the ice—whether crevasses or vertical shafts, which are known to glaciologists as moulins. But what is new is the discovery that as the surface warms, more and more water is pouring into the interior of the ice sheet. Waterfalls as high as 2 miles are taking surface water to the very base of the ice, where it meets the bedrock. "The summer of 2005 broke all records for melting in Greenland," says Hansen. And such melting threatens to destabilize large parts of the ice sheet on timescales measured in years or decades, not millennia.

Jason Box, of Ohio State University, is a young researcher who knows more about this than most. Every year, he visits Swiss Camp, a research station set up in 1990 on Greenland ice. The name was chosen by the camp's founder, Konrad Steffen, of Zurich, so that he felt more at home. The station was originally sited on the equilibrium line, where the ice melt in summer exactly matches the accumulation of new snow in winter. But the equilibrium line has since moved many miles north, as ever-larger chunks of Greenland find themselves in the zone of predominant melting. These days, Box goes boating in an area close to Swiss Camp dubbed the "Greenland Lake District." "Some of these lakes are three or four miles across and have lasted for a decade or more now," he says. "You wouldn't think it was Greenland at all."

The lakes are more than just symptoms of melting. They are also reservoirs for the destruction of the ice sheet. "These lakes keep growing and growing until they find a crevasse, into which they drain," Box says. "Down there are extensive river systems, between the ice and the hard rock, that eventually emerge at the glacier snout. There may be great lakes, too."

Another regular visitor to Swiss Camp is the glaciologist Jay Zwally, one of Hansen's colleagues at NASA. He made the alarming discovery that during warm years the half-mile-thick ice lifts off the bedrock and floats on the water—rising half a yard or more at times. And it floats toward the ocean. Ice sheets are never entirely still, of course. But Swiss Camp is already more than a mile west of where it started. And Zwally found that in summer, when the surface is warmer and more water pours down the crevasses, the velocity of the ice sheet's flow increases. Acceleration starts a few days after the melting begins at the surface. It stops when the melting ceases in the autumn.

This discovery is a revelation, glaciologists admit. "These flows completely change our understanding of the dynamics of ice-sheet destruction," says Richard Alley, of Penn State. "We used to think that it would take 10,000 years for melting at the surface to penetrate down to the bottom of the ice sheet. But if you make a lake on the surface and a crack opens and the water goes down the crack, it doesn't take 10,000 years, it takes ten seconds. That huge lag time is completely eliminated."

As ever, Alley has a good analogy. "The way water gets down to the base of glaciers is rather the way magma gets up to the surface in volcanoes—through cracks. Cracks change everything. Once a crack is created and filled, the flow enlarges it and the results can be explosive. Like volcanic eruptions. Or the disintegration of ice sheets." The lakes on the surface of Greenland are, he says, the equivalent of the pots of magma beneath volcanoes. "More melting will mean more lakes in more places, more water pouring down crevasses, and more disintegration of the ice." No wonder, in a paper in *Science,* Zwally called the phenomenon "a mechanism for rapid, large-scale, dynamic responses of ice sheets to climate warming."

Could such processes be close to triggering a runaway destruction of the Greenland ice sheet? It is hard to be sure, but Greenland does have past form, says David Bromwich, Box's colleague at Ohio State. There is good evidence that the ice sheet lost volume around 120,000 years ago, during the warm era between the last ice age and the previous one. "Temperatures then were very similar to those today," he says. "But the Greenland ice sheet was less than half its present size." He believes that the Greenland ice sheet is a relic of the last ice age whose time may finally have run out. "It looks susceptible, and with the drastic warming we have seen since the 1980s, the chances must be that it is going to melt, and that water will go to the bottom of the ice sheet and lubricate ice flows."

Greenland melting seems to have set in around 1979, and has been accelerating ever since. The interior, above the rising equilibrium line, may still be accumulating snow. But the loss of ice around the edges has more than doubled in the past decade. The NASA team believes that "dynamic thinning" under the influence of the raging flows of meltwater may be responsible for more than half of the ice loss. In early 2006, it reported the results of a detailed satellite radar study of the ice sheet showing that it was losing 180 million acre-feet more of ice every year than it was accumulating through snowfall. That was double the estimated figure for a decade before. And all this gives real substance to the evidence accumulating from Greenland's glaciers, the ice sheet's outlets to the ocean.

Swiss Camp is in the upper catchment of a glacier known as Jakobshavn Isbrae. It is Greenland's largest, flowing west from the heart of the ice sheet for more than 400 miles into Baffin Bay. It drains 7 percent of Greenland. Jakobshavn has for some decades been the world's most prolific producer of icebergs. From Baffin Bay they journey south down Davis Strait; past Cape Farewell, the southern tip of Greenland; and out into the Atlantic shipping lanes. Jakobshavn was the likely source of the most famous iceberg of all—the one that sank the *Titanic* in 1912. But it has been in overdrive since 1997, after suddenly doubling the speed of its flow to the sea. It is now also the world's fastest moving glacier, at better than 7 miles a year.

Jason Box has installed a camera overlooking the glacier to keep track. It takes stereo images every four hours throughout the year. As well as flowing ever faster toward the sea, he says, the glacier is becoming thinner, and in 2003 a tongue of ice 9 miles long that used to extend from its snout into the ocean broke off. "What is most surprising is how quickly this massive volume of ice can respond to warming," says Box. There seems to be a direct correlation between air temperatures in any one year and the discharge of water from glaciers into the ocean. Long time lags, once thought to be a near-universal attribute of ice movement, are vanishing. Jakobshavn, he estimates, could be shedding more than 40 million acre-feet a year, an amount of water close to the flow of the world's longest river, the Nile. Half of that volume is water flowing out to sea from beneath the glacier, and half is calving glaciers.

Other Greenland glaciers are getting up speed, too. The Kangerdlugssuaq glacier, in eastern Greenland, which drains 4 percent of the ice sheet, was flowing into the sea three times faster in the summer of 2005 than when last measured in 1988. At an inch a minute, its movement was visible to the naked eye. Meanwhile, its snout has retreated by three miles in four years. This familiar pattern of faster flow, thinning ice, and rapid retreat of the ice front has also shown at the nearby Helheim glacier, where Ian Howat, of the University of California in Santa Cruz, concludes that "thinning has reached a critical point and begun drastically changing the glacier's dynamics."

Most of these great streams of ice are exiting into the ocean beneath the waterline, in submarine valleys, via giant shelves of floating ice that buttress them. But as the oceans warm, these ice shelves are themselves thinning. It is, says Hansen, a recipe for rapid acceleration of ice loss across Greenland.

The picture, then, is of great flows of ice draining out of Greenland, lubricated by growing volumes of meltwater draining from the surface to the base of the ice sheet and uncorked by melting ice shelves at the coast. All this is new and frightening. "The whole Greenland hydrological system has become more vigorous, more hyperactive," says Box. "It is a very nonlinear response to global warming, with exponential increases in the loss of ice. I've seen it with my own eyes. Even five years ago we didn't know about this." Alley agrees: "Greenland is a different animal from what we thought it was just a few years ago. We are still thinking it might take centuries to go, but if things go wrong, it could just be decades. Everything points in one direction, and it's not a good direction."

"Building an ice sheet takes a long time—many thousands of years," says Hansen. "It is a slow, dry process inherently limited by the snowfall rate. But destroying it, we now realize, is a wet process, spurred by positive feedbacks, and once under way it can be explosively rapid."

8

THE SHELF

Down south, shattering ice uncorks the Antarctic

Over three days in March 2002, there occurred one of the most dramatic alterations to the map of Antarctica since the end of the last ice age. It happened on the shoreline of the Antarctic Peninsula—a tail of mountains 1,200 miles long and more than a mile high pointing from the southern part of the continent toward the tip of South America. A shelf of floating ice larger than Luxembourg and some 650 feet thick, which had been attached to the peninsula for thousands of years, shattered like a huge pane of glass. It broke into hundreds of pieces, each of them a huge iceberg that floated away into the South Atlantic.

There were no casualties, except the self-esteem of Antarctic scientists who believed that after a century of studying the continent's ice, they knew how it behaved. Their subsequent papers revealed their shock. "The catastrophic break-up of the Larsen B ice shelf is remarkable because it reveals an iceberg production mechanism far different from those previously thought to determine the extent of Antarctic ice shelves," wrote Christina Hulbe, a peace activist and glaciologist from Portland State University, in Oregon. Rather than the normal "infrequent shedding of icebergs at the seaward ice front," this time "innumerable icebergs were created simultaneously through the entire breadth of the shelf."

The demise of the Larsen B ice shelf was not in itself a surprise. Both the air and the water around the Antarctic Peninsula had been warming since the 1960s. It had become one of the hot spots of global warming. Warm currents had been gradually eating away at the underside of the floating shelf, while warmer air produced pools of melting water on the surface. It was obvious that the sheet was under strain. Some cracks formed across the surface in 1994; a chunk around the edge of the shelf broke off in 1998. But nothing had prepared glaciologists for what was about to happen. During January 2002, the height of the southern summer, temperatures hit a new high and the heavy winter snow on the shelf's surface began to melt. By the end of the month, satellite pictures showed dark streaks across the shelf. Some were ponds, but others were crevasses that had filled with water.

Water is denser than ice. So, once inside the crevasses, it created pressure that levered them ever wider. There were, in effect, thousands of mechanical wedges pushing ever deeper into the ice shelf. Then, in three climactic days at the start of March, the entire structure gave way. Some 500 billion tons of ice burst into the ocean. In many ways, says Richard Alley, what happened at Larsen B mirrored the processes under way in Greenland. "Water-filled cracks more than a few tens of yards deep can be opened easily by the pressure of water. Ponding of water at the ice surface increases the water pressure wedging cracks open." In their enthusiasm to study ice, glaciologists had forgotten about water.

Larsen B was one of a series of floating shelves formed by ice draining from the mountains of the Antarctic Peninsula. The shelves are the floating front edges of glaciers, and where they meet the ocean, icebergs regularly break off. In recent years, Larsen B had been moving forward by about a yard a day. Despite this constant movement, the ice shelf itself, at more than 650 feet thick, was a surprisingly permanent structure. After its collapse, study of the diatoms in the sediment beneath the former shelf suggested that Larsen B had been there for the entire 12,000 years since the end of the last ice age, when a single ice sheet covered the whole region.

Larsen B wasn't alone; nor has it been alone in disappearing. In all, more than 500 square miles of ice shelves have been lost from around the Antarctic Peninsula in the past half century. The Larsen A ice shelf, the other side of an ice-covered headland called Seal Nunatak, broke up in a storm in 1995. And before that, the Wordie shelf, on the west side of the peninsula, disappeared between 1974 and 1996, triggering a dramatic thinning of the glaciers that fed it. But both were much smaller than Larsen B, and neither disappeared in the catastrophic manner of Larsen B.

"Really we don't think there is much doubt that the collapse of the Larsen B shelf was caused by man-made climate change," says John King, chief climatologist at the British Antarctic Survey (BAS), the inheritor of the great tradition of explorers such as Robert Scott and Ernest Shackleton. From their base at Rothera, on Adelaide Island, BAS researchers have mapped in detail how a pulse of warmer air temperatures has pushed south across the peninsula over the past fifty years, lengthening the summer melt season, sending glaciers into retreat, and destabilizing ice shelves as it goes.

Armed with the evidence of Larsen B, glaciologists are reassessing the stability of dozens of peninsula ice shelves—starting with Larsen C, immediately to the south, which is thinning and widely expected to be the next to go. Eventually, they say, the warming will reach the Ronne ice shelf, a slab of ice the size of Spain at the south of the peninsula. And on the other side of the continent is the Ross ice shelf, the continent's largest. It, too, now seems to be vulnerable, says Hulbe.

Disappearing ice shelves do not contribute to sea level rise because their ice is already floating. Their loss no more raises sea levels than an ice cube melting in a drink causes the glass to overflow. But their disappearance does change what happens inland. Ice shelves buttress the glaciers that feed them. After Larsen B disappeared, it was "as if the cork had been removed from a bottle of champagne," says the French glaciologist Eric Rignot, who works at NASA's Jet Propulsion Laboratory, in California. The glaciers that once discharged their ice onto the Larsen B shelf are now flowing into the sea eight times faster than they did before the shelf collapsed. Similar acceleration has happened after other ice sheet collapses. And that faster discharge of ice from land into the ocean is raising sea levels. With the Ross Sea being the main outlet for several of the largest glaciers on the West Antarctic ice sheet, which contains enough ice to raise sea levels by six yards, the stakes are rising.

9

THE MERCER LEGACY

An Achilles heel at the bottom of the world

John Mercer was an English eccentric and, frankly, somewhat disreputable. The list of charges against him is long. He had a penchant for doing his fieldwork in the nude, and was once convicted for jogging naked near his campus at Ohio State University, in Columbus. He regularly fell out with colleagues, and once abandoned two graduate students, including his acolyte and eventual successor Lonnie Thompson, high in the Andes after the money ran out on a field trip. Thompson thought it was something he'd said, until he realized that "those kinds of things kept happening to John; he was the same with everyone."

Mercer, who died of a brain tumor in 1987, is now a largely forgotten figure outside the glaciology community. But within it he is regarded by many, not least Thompson himself, as a genius. In the late 1940s, he set off alone to explore the ice in distant Patagonia, mapping much of the area, and came to realize that tropical glaciers might hold clues to the history of the world's climate. He is credited with inventing the term "greenhouse effect" during a symposium at Ohio State in the early 1960s. But probably his greatest legacy is in Antarctica, where back in the 1960s he made a prophetic warning that may one day ensure the revival of his memory.

At a time when everyone else saw Antarctic ice as just about the most dependable glacial feature on the planet, Mercer began to argue that much of it may have entirely disintegrated during the last interglacial era, about 125,000 years ago. And, though it took him a decade to get his warning into print, he feared that it might be about to happen again. In 1978, in *Nature,* he published a paper declaring: "I contend that a major disaster—a rapid deglaciation of West Antarctica—may be in progress ... within about 50 years."

The two ice sheets covering Antarctica are vast. The smaller of them, the West Antarctic ice sheet, covers around 1.5 million square miles. It is vulnerable because, unlike its larger eastern neighbor, it does not sit on dry land. Instead, like a giant ship that has foundered in shallows, it is perched precariously on an archipelago of largely submerged mountains. Ocean currents are swirling beneath its giant ice shelves. The sea temperatures today are close to freezing, but the risk is that as they rise, melting will loosen the ice sheet's moorings.

The heart of the West Antarctic ice sheet has some protection from the ocean. On two sides it is buttressed by mountains, and on the other two sides it is held in place by the Ronne and Ross ice shelves. But Mercer warned that if the ice shelves gave way, the entire sheet could lift off and float away: "Climate warming above a critical level would remove all ice shelves, and consequently all ice grounded below sea level, resulting in the deglaciation of most of West Antarctica." Once under way, the disintegration would "probably be rapid, perhaps catastrophically so." Most of the ice sheet would be gone within a century. He reckoned that a warming of 9 degrees would be enough to set the process in train. Parts of the continent have already experienced more than 3.6 degrees of warming. "One warning sign that a dangerous warming is beginning will be the break-up of ice shelves in the Antarctic Peninsula," he said. Like Larsen B.

Another old acolyte of Mercer's is Terry Hughes, of the University of Maine. Back in 1981, he suggested that the West Antarctic ice sheet might have another vulnerability—a "weak underbelly" in Pine Island Bay, a large inlet on the Amundsen Sea, west of the Antarctic Peninsula. This is one of the most remote places on Earth. Head north from Pine Island Bay, and you don't hit land until Alaska. These are dangerous waters—deep, with unusually tall icebergs breaking off the glaciers and being blown fast across the bay by fierce winds. There is a constant danger of getting trapped by the ice if the wind changes. Onshore, the terrain is rugged, and its weather is violent, with intense snowstorms steered inland by the Antarctic Peninsula. Even Antarctic researchers have given Pine Island Bay a wide berth. There are no bases here.

Hughes's "weak underbelly" theory was, like Mercer's warnings a decade before, roundly ignored at the time. When I first wrote about it, a few years later, other glaciologists warned me off, suggesting that it had been discredited. But today, just mentioning Pine Island Bay is enough to send a shudder through the hearts of many glaciologists. Hughes, they now believe, was right on the mark.

The bay is the outlet for two of Antarctica's top five glaciers: Pine Island and Thwaites. Together, they drain about 40 percent of the West Antarctic ice sheet. They were already the fastest-flowing glaciers in Antarctica when, in the 1990s, Pine Island began to accelerate sharply, and Thwaites, while traveling at the same speed, doubled its flow by becoming twice as wide. The glaciers were responding to a rapid melting of their own ice shelves. The melting was in turn caused by warmer seawater circling into the bay.

The discovery of the accelerating glaciers has, once again, turned conventional thinking about the dynamics of ice on its head. The old view holds that events on the coast, where a glacier meets the ocean, have little bearing on what happens inland. But at Pine Island Bay, the impacts of coastal melting are swiftly being felt throughout the glaciers' network of tributaries across the ice sheet. In the past decade, the flow of the two glaciers has speeded up, not just at the coast but for 125 miles inland. The NASA glaciologist Eric Rignot reported in 2004 that the two glaciers are dumping more than 200 million acre-feet of ice a year into Pine Island Bay. This dwarfs even the very heavy snowfall, which adds about 130 million acre-feet a year. The net "mass loss" of ice from the Pine Island Bay catchment has tripled in a decade.

Since Rignot's paper was published, the news has become even grimmer. Studies of the Pine Island glacier show that its ice shelf is thinning fast. As it thins, ever more warm seawater penetrates beneath the glacier. The "grounding line," the farthest point downstream where the ice makes contact with solid rock, has been retreating by more than a mile a year. Once under way, the retreat of the grounding line is "theoretically self-perpetuating and irreversible, regardless of climate forcing," says Rignot. The glacier is primed for runaway destruction.

In 2005, British and Texas researchers flew more than 45,000 miles on more than a hundred flights back and forth across the Pine Island and Thwaites glaciers, using ice-penetrating radar to map the rocks beneath an area of ice the size of France and sometimes nearly 2 miles high. They found that inland along its major tributaries the Pine Island glacier sat on great lakes of meltwater. There seemed to be remarkably little to hold back its flow. Meanwhile, the Thwaites glacier, which is a stream of ice flowing through a wider area of ice sheet, could be about to widen again, says David Vaughan, of the BAS, who masterminded the survey.

If the Pine Island and Thwaites glaciers are on a one-way trip to disaster, the implications are global. Together they drain an area containing enough ice to raise sea levels worldwide by 1–2 yards. In all probability, the Pine Island and Thwaites glaciers are already the biggest causes of sea level rise worldwide. Hughes believes their collapse could destabilize the entire West Antarctic ice sheet, and potentially parts of the East Antarctic ice sheet, too. "The well-documented changes happening just within the past decade are a numbing prospect," he told me. "And we have only hints about exactly what is going on."

Days after Vaughan presented the first findings of the survey to a conference in the U.S., I met Richard Alley. He had been in the audience and had been astounded by the findings. "Thwaites just taps right into the vast reservoirs of ice in the middle of the ice sheet, and the question is whether it will drag them along with it," he said. "I think Thwaites could be absolutely critical. If you pull the plug, the ice goes faster and there is thinning. The only question is whether the plug can re-form a bit further back, or whether the ocean will deliver enough heat for it to just blowtorch its way to the center. I don't think we know the answer to that yet." There was, he said, "a possibility that the West Antarctic ice sheet could collapse and raise sea levels by 6 yards in the next century."

The East Antarctic ice sheet is the biggest, highest slab of ice on the planet. In the unlikely event that it all melted, sea levels would rise by 50 yards or more. But it has been in place for some 20 million years. And in 2005, Curt Davis, of the University of Missouri, reported, after analyzing satellite data, that extra snowfall linked to global warming is raising the height of the ice by almost three quarters of an inch a year—enough to shave current rates of sea level rise by 10 percent. All seemed well, then, with the East Antarctic ice sheet.

But there was a slight problem. Davis's study could cover only the flat interior. Satellite instruments are not yet good enough to establish altitude trends near the coasts, where there is sloping terrain. A footnote to his paper mentions that "mass loss in areas near the coast could be even greater than the gains in the interior." Unfortunately, other researchers say that is precisely what may be happening.

Exhibit A in this case is the Totten glacier. It is a biggie—62 miles wide at its mouth, where it calves icebergs into the Indian Ocean. Totten's network of tributary glaciers drains an area containing more ice than the whole of the West Antarctic. And since the early 1990s, says Andy Shepherd, of the Scott Polar Research Institute, in Cambridge, England, that catchment has been losing enough ice to lower its height by more than 10 yards a year. Another giant of the East Antarctic ice sheet, the Cook glacier, is doing the same.

The last bastion of glacial stability suddenly looks much less safe. And Shepherd points out that Totten and Cook have something else in common with Pine Island, Thwaites, and the other troublesome glaciers on the west side—something suggesting that worse could be ahead. Both Totten and Cook have grounding lines in the ocean that are below sea level—more than 300 yards below in the case of Totten. That is, its contact with the continental land mass is so tenacious that the glacier slides 300 yards under water before the ice gives up contact with the rock and begins to float. That sounds like good news: evidence of stability. The problem is that warmer waters appear to be weakening that contact. Should the grounding line start to retreat, we can expect the glacier to begin the familiar process of thinning and accelerating. The retreat would, in other words, remove the cork from a very large bottle.

Nobody is yet saying that the East Antarctic ice sheet is vulnerable in the way that the western sheet appears to be. It remains very big and, by and large, extremely stable. But, as Rignot puts it, "it is not immune." And every new discovery seems to raise the stakes for the fate of the Antarctic ice. As recently as 2001, the IPCC reported a scientific consensus that it was "very unlikely" that Antarctica would produce any significant rise in sea levels during the twenty-first century. Few glaciologists are repeating that claim with any confidence now. Most would agree with Alley that "major changes are taking place in the Antarctic, on much shorter time scales than previously anticipated."

The British Antarctic Survey now employs a mathematician full time to apply chaos and complexity theory to the fate of the continent's ice—a topic once considered to be of the utmost simplicity. The BAS is using the language of fractals, phase space, and bifurcations to work out what might happen next to the ice sheets of the Antarctic Peninsula and the glaciers of Pine Island Bay. Its scientists have seen Larsen B shatter in three days; they believe they are seeing the soft underbelly of the West Antarctic ice sheet ripped open before their eyes. What next?

RISING TIDES

Saying "toodle-oo" to Tuvalu

The Carteret Islands are to be abandoned. Life is simply too hard for their 2,000 inhabitants, huddled on a clutch of low-lying coral islands in the South Pacific, with a total surface area of just 150 acres, and rising sea levels threatening to wash them away. The islands, named after an eighteenth-century English explorer of the South Seas, Philip Carteret, have been under nearly constant erosion since the 1960s, and the current guess is that they will be wholly submerged by 2015. Already their fields have been invaded by salt water, and the breadfruit crops have died. The people, refugees in their own land, depend on handouts.

In 2001, when strong winds and rough seas cut off the atoll and prevented them from going to sea to catch fish, many resorted to eating seaweed. One resident on the island of Han pleaded by radio for rescue: "Erosion is occurring from both sides, and the island is getting narrow. In Piul, many families are leaving. Huene Island is divided in half, and four families only are left. On Iolasa, Iosela, and Iangain, when high seas occur, they stand below sea level. This is very frightening." Indeed. In November 2005, the central government in Papua New Guinea, of which the Carteret Islands form a part, agreed that the islanders should all be moved to Bougainville, a four-hour boat ride to the southwest. Ten families at a time will journey over the next few years, relinquishing their ancestral homes forever.

For most people around the world, stories of a rise in sea level remain a matter of academic interest, if that. The risks seem remote. But for the inhabitants of low-lying islands like the Carterets, it is happening now and devastating their lives.

The 10,000 citizens of the nine inhabited South Pacific islands of Tuvalu are also abandoning ship. High tides regularly wash across the main street in the capital, Funafuti; sea salt is poisoning their fields and killing their coconuts. Tuvalu is a full-fledged nation-state. Formerly the British Ellice Islands, it won independence in 1975. But just thirty years on, it seems destined to be the first modern nation-state to disappear beneath the waves. A twenty-first-century Atlantis. "In fifty years, Tuvalu will not exist," says the prime minister. His government has signed a deal with New Zealand, 1,800 miles away, that will allow the entire population to move there in the coming years, as rising tides and worsening storms destroy their homes.

One by one, the island nations of the South Pacific are drowning. Kiribati, formerly the British Gilbert Islands, won its independence on the same day as Tuvalu. It, too, is going under. Two uninhabited islands disappeared in 1999. The following year, Nakibae Teuatabo, a resident of Kiribati, explained its plight to me at a climate-change conference in Bonn, where he had been sent to plead for his country's survival. "Eight or nine house plots in the village that my family belongs to have been eroded. I remember there was a coconut tree outside the government quarters where I lived. Then the beach all around it was eroded, and eventually the tree disappeared. It might not sound a lot to you. But the atolls are just rings of narrow islands surrounding a lagoon, with the open ocean on the outside. Some of the islands are only a few yards wide in places. Imagine standing on one of these islands with waves pounding on one side and the lagoon on the other. It's frightening."

Villagers on some outer islands have already moved away as the sea gobbles up their land, he said. "Apart from causing coastal erosion, higher tides are pushing salt water into the fields and into underground freshwater reservoirs. In some places, it just bubbles up from the ground." It was a heart-rending story—good for journalists, but of no interest to most government negotiators at the conference. Such nations, it seems, are expendable.

The world's sea levels have been largely stable for the past 5,000 years, since the main phase of melting of ice sheets after the end of the last ice age abated. Some residual ice loss continued to raise sea levels at less than one hundredth of an inch a year. But around 1900, the rise began to increase. At first, this was most likely owing to the melting of glaciers after the little ice age ended, in the mid-nineteenth century. That should have diminished during the twentieth century. But instead it has accelerated in the past fifty years, to around 0.08 inches a year. About half of this increase is probably due to the process known to physicists as thermal expansion. And the rest is probably due to the resumed melting of the world's glaciers and ice caps, doubtless largely a result of man-made climate change.

The first signs of a further acceleration emerged in the early 1990s, when satellite data suggested a sudden rise of 0.11 inches a year. Since 1999, it may have risen further, to 0.14 inches. At the time of this writing, these figures had failed to gain much attention, because glaciologists remained worried about their reliability. Some think there may be a problem calibrating the satellite data; others that it may simply be a natural fluctuation. But, with every year that passes, more researchers are concluding that we are seeing the first effects of the dramatic changes apparently under way on the ice sheets of Greenland and Antarctica.

The planet has a history of startling sea level rise that cannot be explained by the conventional models used by glaciologists to predict future change. Consider events toward the end of the last ice age. Around 20,000 years ago, at what glaciologists call the "glacial maximum," so much water was tied up in ice on land that sea levels were around 400 feet lower than they are today. Then a thaw began. Sea levels initially rose by around 0.4 inches a year. That is four or five times faster than today, but within the traditional expectations of glaciologists. Then something happened. About 14,500 years ago, the tides went haywire. Within 400 years, sea levels rose by 65 feet. That's an average rate of just over a yard every twenty years.

It is worth thinking about those numbers. If such a rise happened today, you could say "toodle-oo" to Tuvalu by 2010; most of Bangladesh would be under water by 2020; millions of people on the Nile Delta would be looking for new homes by 2025; London would need a new Thames Barrier immediately. New Orleans? Well, forget New Orleans, and Florida, and most of the rest of the U.S. seaboard, too. Lagos, Karachi, Sydney, New

York, Tokyo, Bangkok: you name your coastal megacity, and it would be abandoned by midcentury. It sounds unbelievable, but we know the rise happened. The evidence is in tidemarks on ancient cliffs and in the remains of coral that can live only close to sea level.

How could such a thing have happened? It required the transfer into the oceans of about 13 billion acre-feet of ice every year throughout the 400-year period. That is a huge amount of ice. Glaciologists believe that the West Antarctic ice sheet, which was much larger then, was the most likely source. But wherever it came from, it could have reached the oceans in such quantities and at such speed only by some process in addition to melting. Such discharges required the physical collapse of ice sheets on a grand scale. That can have happened only if the ice sheets were lubricated at their base by great rivers of meltwater, and destabilized at the coasts by the shattering of ice shelves.

Go back further. In the last interglacial period, about 120,000 years ago, evidence such as wave-cut notches along cliffs in the Bahamas show sea levels 20 feet higher than they are today. During a previous interglacial, some 400,000 years ago, they may have been even higher. In neither period were temperatures significantly higher than they are today. On the face of it, either the West Antarctic ice sheet, or the Greenland ice sheets, or both, succumbed at temperatures close to our own. We can expect that temperatures will rise by about 3 to 5 degrees within the coming century. That, says Hansen, would make them as high as they were 3 million years ago, before the era of ice ages started. What were sea levels then? About 25 yards higher than today, plus or minus 10 yards, he says.

A first guess is that we will very soon have set the world on a course for reaching such levels again. The models of glaciologists suggest that, if this happens, it will take thousands of years. Jim Hansen doesn't believe it. "I'm a modeler, too, but I rate data higher than models," he says. He already sees evidence of the start of runaway melting in Greenland and Antarctica, and anticipates that "sea levels might rise by a couple of yards this century, and several more the next century."

Some see this prognosis as alarmist. Where, they say, is the evidence of big sea level rises so far? Hansen says that much of the extra melting has been camouflaged by increased snowfall on the ice sheets: "Because of this, sea level changes slowly at first, but as global warming gets larger, as summer melt extends higher up the ice sheet, and as buttressing ice shelves melt away, multiple positive feedbacks come into play, and the nonlinear disintegration wins the competition, hands down."

The world's ice sheets are "a ticking time bomb," he says. There is no reason why the events of 14,000 years ago should not be repeated in the twenty-first century. "The current planetary energy imbalance is now pouring energy into the Earth system at a rate sufficient to fuel rapid deglaciation." Hansen's hunch is that an increasing amount of global warming will be harnessed to melting the ice sheets. That could slow the heating of the atmosphere, but at the price of faster-rising sea levels. Within a few decades, vast armadas of icebergs could be breaking off the Greenland ice sheet, making shipping lanes impassable and cooling ocean surfaces like the ice in a gin and tonic. Sea level rise, he concludes, is "*the* big global issue." He believes it will transcend all others in the coming century.

It is easy to forget the plight of the people of the Carteret Islands and Tuvalu. Few of us could even find these places on the map. But as the tides rise ever higher, and as the precarious state of the big ice sheets becomes more apparent, we might want to heed those people's fate. It could be that of our own children.

III

Riding the carbon cycle

IN THE JUNGLE

Would we notice if the Amazon went up in smoke?

The Amazon rainforest is the largest living reservoir of carbon dioxide on the land surface of Earth. Its trees contain some 77 billion tons of carbon, and its soils perhaps as much again. That is about *twenty* years' worth of man-made emissions from burning fossil fuels. The rainforest is also an engine of the world's climate system, recycling both heat and moisture. More than half of the raindrops that fall on the forest canopy never reach the ground; instead they evaporate back into the air to produce more rain downwind. The forest needs the rain, but the rain also needs the forest.

But as scientists come to understand the importance of the Amazon for maintaining climate, they are also discovering that it may itself be under threat from climate change. We are familiar enough with the damage done to the world's biggest and lushest jungle by farmers armed with chain saws and firebrands. But, hard as they try, they can destroy the rainforest only slowly. Despite many decades of effort, most of this jungle, the size of western Europe, remains intact. Climate change, on the other hand, could overwhelm it in a few years.

Until recently, many ecologists have thought of the Amazon rainforest much as their glaciologist colleagues conceived of the Greenland ice sheet: as big and extremely stable. The Greenland ice maintained the climate that kept the ice securely frozen, while the Amazon rainforest maintained the rains that watered the forest. But, just as with the Greenland ice sheet, the idea that the Amazon is stable has taken a knock: some researchers believe that it is in reality a very dynamic place, and that the entire ecosystem may be close to a tipping point beyond which it will suffer runaway destruction in an orgy of fire and drought. Nobody is quite sure what would happen if the Amazon rainforest disappeared. It would certainly give an extra kick to climate change by releasing its stores of carbon dioxide. It would most likely diminish rainfall in Brazil. It might also change weather systems right across the Northern Hemisphere.

One man who is trying to find out how unstable the Amazon rainforest might be is Dan Nepstad, a forest ecologist nominally attached to the Woods Hole Research Center, in Massachusetts, but based for more than two decades in the Amazon. He doesn't just watch the forest: he conducts large experiments within it. In 2001, Nepstad began creating a man-made drought in a small patch of jungle in the Tapajos National Forest, outside the river port of Santarem. Although in most years much of the Amazon has rain virtually every day, Tapajos is on the eastern fringe of the rainforest proper, where weather cycles can shut down the rains for months. The forest here is, to some extent, adapted to drought. But there are limits, and Nepstad has been trying to find out where they lie.

He has covered the 2.5-acre plot with more than 5,000 transparent plastic panels, which let in the sunlight but divert the rain into wooden gutters that drain to canals and a moat. Meanwhile, high above the forest canopy, he has erected gantries linked by catwalks, so that he can study the trees in detail as the artificial drought progresses. The work was all done by hand to avoid damaging the dense forest, and the scientists soon found they were not alone. The canals became "congregating places for every kind of snake you can imagine," says Nepstad. Caimans and jaguars cruised by, just, it seemed, to find out what was going on.

The results were worth the effort. The forest, it turns out, can handle two years of drought without great trouble. The trees extend their roots deeper to find water and slow their metabolism to conserve water. But after that, the trees start dying. Beginning with the tallest, they come crashing down, releasing carbon to the air as they rot, and exposing the forest floor to the drying sun. By the third year, the plot was storing only about 2 tons of carbon, whereas a neighboring control plot, on which rain continued to fall, held close to 8 tons. The "lock was broken" on a corner of one of the planet's great carbon stores. The study shows that the Amazon is "headed in a terrible direction," wrote the ecologist Deborah Clark, of the University of Missouri, discussing the findings in *Science*. "Given that droughts in the Amazon are projected to increase in several climate models, the implications for these rich ecosystems are grim."

Everywhere in the jungle, drought is followed by fire. So, in early 2005, Nepstad started an even more audacious experiment. He set fire to another stretch of forest with kerosene torches. "We want to know if recurring fire may threaten the very existence of the forest," he says. The initial findings were not good: the fires crept low along the forest floor, and no huge flames burst through the canopy. The fire may even have been invisible to the satellites that keep a constant watch overhead. But many trees died nonetheless, as their bark scorched and the flow of sap from their roots was stanched.

Nepstad's experiments are part of a huge international effort to monitor the health of the Amazon, called the Large-scale Biosphere-Atmosphere Experiment in Amazonia. From planes and satellites and gantries above the jungle, researchers from a dozen countries have been sniffing the forest's breath and assessing its survival strategies. The current estimate is that fires in the forest are releasing some 200 million tons of carbon a year—far more than is absorbed by the growing forest. The Amazon has become a significant source of carbon dioxide, adding to global warming. More worrying still, the experiment is discovering a drying trend across the Amazon that leaves it ever more vulnerable to fires. Nepstad's work suggests that beyond a certain point, the forest will be unable to recover from the fires, and will begin a process of rapid drying that he calls the "savannization" of the Amazon.

And even as he concluded his drought experiment, nature seemed to replicate it. The rains failed across the Amazon through 2005, killing trees, triggering fires, and reducing the ability of the forest to recycle moisture in future—thus increasing the risk of future drought. Nepstad's experiments suggest that the rainforest is close to the edge—to permanent drought, rampant burning, savannization, or worse. In the final weeks of 2005, the rains returned. The forest may recover this time. But if future climate change causes significant drying that lasts from one year to the next, feedbacks in the forest could realize Nepstad's worst fears.

The 2005 drought was caused by extremely warm temperatures in the tropical Atlantic—the same high temperatures that are believed to have caused the record-breaking hurricane season that year. The rising air that triggered the hurricanes eventually came back to earth, suppressing the formation of storm clouds over the Amazon. And, as I discovered at Britain's Hadley Centre for Climate Prediction, that is precisely what climate modelers are forecasting for future decades.

The Hadley Centre's global climate model is generally regarded as one of the world's top three. And it predicts that business-as-usual increases in industrial carbon dioxide emissions worldwide in the coming decades will generate warmer sea temperatures, subjecting the Amazon to repeated droughts, and thus creating "threshold conditions" beyond which fires will take hold. The Amazon rainforest will be dead before the end of the century. Not partly dead, or sick, but dead and gone. "The region will be able to support only shrubs or grass at most," said a study published by the Hadley Centre in 2005.

Not all models agree about that. But the Hadley model is the best at reproducing the current relationship between ocean temperatures and Amazon rainfall, so it has a good chance of being right about the future, too. Nepstad himself predicts that a "megafire event" will spread across the region. As areas in the more vulnerable eastern rainforest die, they will cease to recycle moisture back into the atmosphere to provide rainfall downwind. A wave of aridity will travel west, creating the conditions for fire to rip through the heart of the jungle.

With the trees gone, the thin soils will bake in the sun. Rainforest could literally turn to desert. The Hadley forecast includes a graph of the Amazon's forest's future carbon. It predicts that the store of a steady 77 billion tons over the past half century will shrink to 44 billion tons by 2050 and 16.5 billion tons by the end of the century. That, it calculates, would be enough to increase the expected rate of warming worldwide by at least 50 percent.

The Amazon rainforest does not just create rain for itself. By one calculation, approaching 6 trillion tons of water evaporates from the jungle each year, and about half of that moisture is exported from the Amazon basin. Some travels into the Andes, where it creates clouds that swathe some mountains so tightly that their surfaces have never been seen by satellite. Some blows south to water the pampas of Argentina, some east toward South Africa, and some north toward the Caribbean. The forest is a vital rainmaking machine for most of South America. As much as half of Argentina's rain may begin as evaporation from the Amazon.

But the benefits of the great Amazonian hydrological engine extend much further, and are not restricted to rainfall. The moisture also carries energy. A lot of solar energy is used to evaporate moisture from the forest canopy. This is one reason why forests stay cooler than the surrounding plains. And when the moisture condenses to form new clouds, that energy is released into the air. It powers weather systems and high-level winds known as jets far into the Northern Hemisphere. Nicola Gedney and Paul Valdes, two young climate researchers at the University of Reading, have calculated that this process ultimately drives winter storms across the North Atlantic toward Europe. "There is a relatively direct physical link between changes over the deforested region and the climate of the North Atlantic and western Europe," they say. If the rainforest expires, the hydrological engine, too, is likely to falter, and the link will be cut.

12

WILDFIRES OF BORNEO

Climate in the mire from burning swamp

The smoke billowed through Palangkaraya. One of the largest towns in Borneo was engulfed in acrid smog denser even than one of London's old pea-soupers. It blotted out so much sun that there was a chill in the air of a town more used to the dense, humid heat of the rainforest that encircled it. This was late 1997, and the rainforest was burning. The most intense El Niño event on record in the Pacific Ocean had stifled the storm clouds that normally bring rain to Borneo and the other islands of Indonesia. Landowners took advantage of the dry weather to burn the forest and carve out new plantations for palm oil and other profitable crops. The fires got out of control, and the result was one of the greatest forest fires in human history. The smoke spread for thousands of miles. Unsighted planes crashed from the skies, and ships collided at sea; in neighboring Malaysia and distant Thailand, hospitals filled with victims of lung diseases, and schools were closed. The fires became a global news story. The cost of the fires in lost business alone was put at tens of billions of dollars.

But it was not just the trees that were burning. The densest smoke was in central Borneo, around Palangkaraya, where the fires had burrowed down, drying and burning a vast peat bog that underlay the forest. The peat, 60 feet deep in many places, was the accumulated remains of wood and forest vegetation that had fallen into the swamps here over tens of thousands of years. Even after the rains returned, the peat continued to smolder for months on end. When the smoke finally cleared, most of the swamp forest was burned and black, and skeletons of trees poked from charred ground that had shrunk in places by a yard or more.

The burning of the Borneo swamp was part of a wider global assault on tropical rainforests—for timber and for land. But there were aggravating factors here. Until recently, the swamps were empty of humans. Local tribes and modern farmers alike had found them inhospitable and inaccessible. But in the early 1990s, Indonesia's President Suharto decreed that an area of the central Borneo swamp forest half the size of Wales should be drained and transformed into a giant rice paddy to make his country self-sufficient in its staple foodstuff. Some 2,500 miles of canals were dug to drain the swamp. Some 60,000 migrant farmers were brought in from other islands to cultivate the rice. The soils proved infertile, and virtually no rice was ever grown. The megaproject was abandoned. But its legacy lingers, as the canals continue to drain the swamps, and the desiccated peat burns every dry season. Especially during El Niños.

This is no mere local environmental disaster. Jack Rieley, a British ecologist with a love of peat bogs who has adopted the central Borneo swamps for his field studies, says the disaster is of global importance. At least half of the world's tropical peat swamps are on the Indonesian islands of Borneo, Sumatra, and West Papua. And the largest, oldest, and deepest of them are in central Borneo, where they cover an area a quarter the size of England and harbor large populations of sun bears and clouded leopards, as well as the world's largest surviving populations of orangutans. They also contain vast amounts of carbon—perhaps 50 billion tons of the stuff. That is almost as much as in the entire Amazon rainforest, which is more than ten times as large. One acre of Borneo peat swamp contains 880 tons of carbon.

Tropical peat swamps are a major feature of the planet's carbon cycle. They are important amplifiers of climate change, capable of helping push the world into and out of ice ages by capturing and releasing carbon from the air. For thousands of years, they have been keeping the world cooler than it might otherwise be, by soaking up carbon from the air. For that carbon to be released now, as the world struggles to counter global warming, would be folly indeed. But that is what is happening. Rieley estimates that during the El Niño event of 1997 and 1998, as Palangkaraya disappeared for months beneath smoke, the smoldering swamps lost more than half a yard of peat layer, and released somewhere between 880 million and 2.8 billion tons of carbon into the atmosphere: the equivalent of up to 40 percent of all emissions from burning fossil fuels worldwide that year.

At first there was some skepticism about his figures. Few other researchers had been to Borneo to see what was going on. But in 2004, U.S. government researchers published a detailed analysis of gas measurements made around the world. It showed that roughly 2.2 billion tons more carbon than usual entered the atmosphere during 1998—and two thirds of that excess came from Southeast Asia. The Borneo fires must have contributed most of that, and burning peat was almost certainly the major component. "We are witnessing the death of one of the last wilderness ecosystems on the planet, and it is turning up the heat on climate change as it goes," says Rieley. "What was once one of the planet's most important carbon sinks is giving up that carbon. The whole world is feeling the effect."

Every year, farmers continue burning forest in Borneo to clear land for farming. And whenever the weather is dry, those fires spread out through the jungle and down into the peat. Satellite images suggest that 12 million acres of the swamp forests were in flames at one point during late 2002. And 2002 and 2003 were the first back-to-back years in which net additions to the atmosphere's carbon burden exceeded 4.4 billion tons. Rieley reckons that the burning swamp forests contributed a billion tons of that.

It looked as if smoldering bogs in remote Borneo were single-handedly ratcheting up the speed of climate change. They show, says David Schimel, of the National Center for Atmospheric Research (NCAR), in Boulder, Colorado, how "catastrophic events affecting small areas can have a huge impact on the global carbon balance." Fire in Borneo and the Amazon may be turning the world's biggest living "sinks" for carbon dioxide into the most dynamic new source of the gas in the twenty-first century.

13

SINK TO SOURCE

Why the carbon cycle is set for a U-turn

It seemed too good to be true. Throughout the 1980s and 1990s, evidence grew that wherever forests survived around the world, they were growing faster. And as they did so, they were soaking up ever more carbon dioxide from the air. Despite deforestation in the tropics, the world's forests overall were a strong carbon sink. Most researchers assumed that the extra growth happened because rising concentrations of carbon dioxide in the atmosphere made it easier for trees to absorb the gas from the air. Provided that the other ingredients for photosynthesis, such as water and nutrients, were available, the sky was the limit for growing plants. The "CO_2 fertilization effect" entered the climate scientists' lexicon.

In 1998, at the height of this enthusiasm, a group of carbon modelers at Princeton University scored what looked like a political as well as a scientific bull's-eye. Song-Miao Fan and colleagues claimed in a paper in *Science* to have discovered "a large terrestrial carbon sink in North America." The U.S. and Canada, they said, had become a hot spot for carbon absorption, as trees grew on abandoned farmland and previously logged forests, and carbon dioxide in the air boosted growth. They calculated the sink at a stunning 2.2 billion tons a year—more than enough to offset the two countries' total annual emissions from power plants, cars, and the rest. Thanks to their trees, the biggest polluters on the planet were "carbon neutral."

To many, it seemed an outrageous claim. And on examination, it turned out to involve some fairly heroic assumptions about where carbon dioxide in the atmosphere was coming from, where it was going, and how it moved around. Carbon-cycle specialists poured cold water on the notion. The figure of 2.2 billion tons was not far off the total amount of carbon that North America's trees absorbed in a year in order to grow. If it was accurate, it meant that no North American trees were dying; they weren't even breathing out—because both processes release carbon dioxide back into the air. But the findings came less than a year after the Clinton administration had signed up for tough carbon-dioxide-emissions targets at Kyoto, without any clear idea of how it was going to achieve them. They seemed like manna from heaven.

And yes, it was too good to be true. The authors agreed that their data were sparse and their analytical techniques largely untried. Nobody, it turned out, could repeat the results. A plethora of researchers demonstrated that U.S. forests could never have stashed away more than a fifth of the nation's emissions. After a while, nobody stood up to defend the original results, and they disappeared from view as fast as they had arrived.

The final nail entered the coffin when it emerged later that 1998, when the report was published, was just about the worst year on record for nature's ability to soak up carbon dioxide from the air. Forests and peat bogs had burned from the Amazon to Borneo. If there had been a big sink, it was disappearing even as it was uncovered. And it wasn't just in the tropics that carbon had been seeping out of the biosphere. There were major forest fires from Florida to Sardinia, and from Peru to Siberia—where Russian foresters revealed that a conflagration on a par with Borneo's had been taking place virtually unnoticed. The world's largest stretch of forest, which for 200 years had been soaking up a fraction of Europe's industrial emissions as they poured east on the prevailing winds, was giving up what it had previously absorbed. As 1998 closed, the idea of a huge carbon sink in the U.S. or anywhere else seemed absurd.

The next episode in the story of the amazing disappearing carbon sink came in the summer of 2003, when Europe suffered a massive heat wave. Temperatures averaged 10°F above normal during July. In France, the mercury soared above 104°F. With the high temperatures accompanied by less than half the usual rainfall, Europe's beech trees and cornfields, grasslands and pine forests, were expiring.

Philippe Ciais, a Paris-based environmental scientist, followed events.

He was a key player in CarboEurope, a project begun a couple of years earlier to measure Europe's carbon sink. It was launched in the aftermath of the purported discovery of the large North American carbon sink. European politicians, like their U.S. counterparts, were keen to discover if nature was helping them meet their own Kyoto Protocol targets. Ciais's initial assessment was that, thanks to warmer temperatures, higher carbon dioxide levels in the air, and a longer growing season, Europe's ecosystems were absorbing up to 12 percent of its man-made emissions.

But in 2003, the carbon sink blew a fuse. During July and August that year, when Europe's ecosystems would normally have been in full bloom and soaking up carbon dioxide at their fastest, around 550 million tons of carbon escaped from western European forests and fields. This was roughly equivalent to twice Europe's emissions from burning fossil fuels during those two months. All the carbon absorbed in recent years was being dumped back into the atmosphere in double-quick time. The rapid exhaling of the continent's ecosystems was "unprecedented in the last century," said Ciais. But he judged that it was likely to be repeated "as future droughts turn temperate ecosystems from carbon sinks into carbon sources."

Europe seemed to have fast-forwarded into a nightmare future strapped to a runaway greenhouse effect. And it soon emerged that Europe's carbon crisis was part of a more general story of summer stress across the Northern Hemisphere. Ning Zeng, of the University of Maryland, found an area of drought stretching from the Mediterranean to Afghanistan. It had lasted from 1998 to 2002, and had eliminated a natural carbon sink across the region that had averaged 770 million tons a year over the previous two decades.

Alon Angert, of the University of California at Berkeley, explained the big picture. Through the 1980s and into the early 1990s, the "CO_2 fertilization effect" had been working rather well, with increased photosynthesis in the Northern Hemisphere soaking up ever more carbon dioxide. But sometime around 1993 that had tailed off, probably because of droughts and higher temperatures. And since the mid-1990s, the carbon sink had been in sharp decline. From the Mediterranean to central Asia, and even in the high latitudes of Siberia and northern Europe, the added uptake of carbon by plants in the early spring was canceled out by the heat and water stress of hotter, drier summers. The findings, Angert said, dashed widespread expectations of a continuing "greening trend" in which warm summers would speed plant growth and moderate climate change. Instead, "excess heating is driving the dieback of forests, accelerating soil carbon loss and transforming the land from a sink to a source of carbon to the atmosphere."

And further north, beyond the tree line, where some of the fastest warming rates in the world are currently being experienced, fear is growing about the carbon stored in the thick layers of permanently frozen soil known as permafrost. The carbon comprises thousands of years' accumulation of dead lichen, moss, and other vegetation that never had a chance to rot before it froze. David Lawrence, of the NCAR, reported in 2005 that he expected the top 3 yards of permafrost across most of the Arctic to melt during the twenty-first century. This will leave a trail of buckled highways, toppled buildings, broken pipelines, and bemused reindeer; it will also unfreeze tens and perhaps hundreds of billions of tons of carbon. As the thawed vegetation finally rots, most of its carbon will return to the atmosphere as carbon dioxide. In those bogs and lakes where there is very little oxygen, most of the carbon will be converted into methane—which, as we will see in the next chapter, is an even more potent greenhouse gas.

We should not write off the carbon sink entirely. It won't die altogether. Especially in higher latitudes, warmer and wetter conditions will sometimes mean that trees grow faster and farther north than before—at least where plagues of insects don't get them first. Right now, the best guess is that, on average, forests are still absorbing more carbon dioxide than they release. Up to a fifth of the carbon dioxide emitted by burning fossil fuels may still be being absorbed by soils and forests. But the sink is diminishing, not rising as many anticipated. And many believe that the sink is doomed as we face more and more years like 1998 and 2003.

One of those who fear the worst is Peter Cox, a top young British climate modeler who left the Hadley Centre to spend more time investigating the carbon cycle at the Centre for Ecology and Hydrology, at Winfrith, in Dorset. He believes he is on the trail of the disappearing carbon sink, and is prepared to put a date on when it will disappear. "Basically, we are seeing two competing things going on," he says. "Plants absorb carbon dioxide as they grow through photosynthesis; but they give back the carbon dioxide as they die and their wood, leaves, and roots decompose. The speed of both processes is increasing."

First, the extra carbon dioxide in the atmosphere encourages photosynthesis to speed up. So plants grow faster and absorb more carbon dioxide. But that extra carbon dioxide is also warming the climate. And the warming encourages the processes that break down plant material and release carbon dioxide back into the air. Because it takes a couple of decades for the extra carbon dioxide to bring about warmer temperatures, we have seen the fertilization effect first. Now the process of decay is starting to catch up.

The processes do not involve plants alone. Soils have their own processes of inhaling and exhaling carbon. And they, too, will switch from being a net sink to a net source—eventually releasing what carbon they have absorbed in recent decades. Ultimately, "you can't have the one without the other," Cox says. "If you breathe in, eventually you have to breathe out." And soon, most of the rainforests and soils of the world will be breathing out, pouring their stored carbon back into the air. If the climate gets drier and more fires occur, then the release of the carbon dioxide will happen even more quickly. But it will happen anyway.

The entire land biosphere—the forests and soils and pastures and bogs—has been slowing the pace of global warming for some decades. Soon the biosphere will start to speed it up. The day the biosphere turns from sink to source will be another tipping point in Earth's system. Once under way, the process, like collapsing ice sheets, will be unstoppable. Potentially, hundreds of billions of tons of carbon in the biosphere could be destabilized, says Pep Canadell, a carbon-cycle researcher for the Australian government research agency CSIRO.

Nobody is quite sure when the tipping point might occur. "It is possible," says Cox, "that the 2003 surge of carbon dioxide into the atmosphere is the first evidence." But while some parts of the biosphere may now be irrevocably stuck as carbon sources, the entire system is likely to take a few decades to switch. But of course, much will probably depend on how fast we allow temperatures to rise.

Cox suggests that 2040 is probably when the biosphere will start taking its revenge on us for relying on its accommodating nature. He calculates that by the end of the century, the biosphere could be adding as much as 8 billion tons of carbon to the atmosphere each year. That is roughly the amount coming each year from burning fossil fuels today, and probably enough to add an extra 2 or 3°F to global temperatures—degrees that are not yet included in the IPCC forecasts.

Only one country, so far as I am aware, has completed anything like a national study of the current impact of these changes on its carbon budget. Perhaps understandably, such studies have a lower priority since nature was shown to be unlikely to offer a helping hand in meeting Kyoto targets. But Guy Kirk, of the National Soil Resources Institute, part of Cranfield University, has done the job for Britain. He surveyed 6,000 test plots across forest and bog, heath and farmland, scrub and back gardens, to see how much carbon dioxide is leaving the biosphere and how much is entering it. His conclusion is that the British biosphere is releasing about 1 percent of its carbon store into the atmosphere every year. Enough, in other words, to turn the whole country into desert in one century.

    Kirk rules out altered methods of farming or land use as the predominant cause. The increase is so universal that it can only be owing to climate change. He puts the national release at around 14 million tons a year. That, he points out, is roughly the amount of carbon dioxide the British government has kept from the atmosphere each year in its efforts to comply with the Kyoto Protocol. As the German researcher Ernst-Detlef Schulze, of CarboEurope, puts it—rather gloatingly, I think—this "completely offsets the technological achievements of reducing carbon dioxide emissions, putting the UK's success in reducing greenhouse gas emissions in a different light." True enough. But Britain is not alone.

14

THE DOOMSDAY DEVICE

A lethal secret stirs in the permafrost

One of my favorite films is *Dr. Strangelove.* It was made back in 1964, when the biggest global threat was nuclear Armageddon. Directed by Stanley Kubrick, and starring Peter Sellers as Dr. Strangelove, a wheelchair-bound caricature of Henry Kissinger, the film was a satire of the military strategy known as Mutual Assured Destruction—or MAD, for short. The plot involved the Soviet Union's building the ultimate defense, a doomsday device in the remote wastes of Siberia. If Russia were attacked, the device would shroud the world in a radioactive cloud and destroy all human and animal life on earth. Unfortunately, the Soviet generals forgot to tell the Americans about this, and, needless to say, Dr. Strangelove and the American military attacked. The film ends with a deranged U.S. officer (played by Slim Pickens) sitting astride a nuclear bomb as it is released into the sky above Siberia. The end of the world is nigh, as the credits roll.

Now our most feared global Armageddon is climate change. But reason to fear truly does lurk in the frozen bogs of western Siberia. There, beneath a largely uninhabited wasteland of permafrost, lies what might reasonably be described as nature's own doomsday device. It is primed to be triggered not by a nuclear bomb but by global warming. That device consists of thick layers of frozen peat containing tens of billions of tons of carbon.

The entire western Siberian peat bog covers approaching 400,000 square miles—an area as big as France and Germany combined. Since its formation, the moss and lichen growing at its surface have been slowly absorbing massive amounts of carbon from the atmosphere. Because the region is so cold, the vegetation only partially decomposes, forming an ever-thickening frozen mass of peat beneath the bog. Perhaps a quarter of all the carbon absorbed by soils and vegetation on the land surface of Earth since the last ice age is right here.

The concern now is that as the bog begins to thaw, the peat will decompose and release its carbon. Unlike the tropical swamps of Borneo, which are degrading as they dry out, and producing carbon dioxide, the Siberian bogs will degrade in the wet as the permafrost melts. In fetid swamps and lakes devoid of oxygen, that will produce methane. Methane is a powerful and fast-acting greenhouse gas, potentially a hundred times more potent than carbon dioxide. Released quickly enough in such quantities, it would create an atmospheric tsunami, swamping the planet in warmth. But we have to change tense here. For "would create," read "is creating."

In the summer of 2005, I received a remarkable e-mail from a man I had neither met nor corresponded with, a young Siberian ecologist called Sergei Kirpotin, of Tomsk State University, in the heart of Siberia. A collaborator of his at Oxford University had suggested me as a Western outlet for what Kirpotin in his e-mail called an "urgent message for the world." He had recently undertaken an expedition across thousands of miles of the empty western Siberian peatlands between the bleak windswept towns of Khatany-Mansiysk, Pangody, and Novy Urengoi. Nobody, barring a few reindeer herders, lives out here. It was an area that Kirpotin and his colleagues had visited several times in the past fifteen years, observing the apparently unchanging geography and biology of the tundra. This time they had found a huge change.

"We had never seen anything like it, and had not expected it," he said. Huge areas of frozen peat bog were suddenly melting. The former soft, spongy surface of lichens and moss was turning into a landscape of lakes that stretched unbroken for hundreds of miles. He described it as an "ecological landslide that is probably irreversible and is undoubtedly connected to climate warming." Most of the lakes had formed, he said, since his previous visit, three years before. There was clearly a huge danger that the melting peatland would begin to generate methane.

I had come across Russian scientists before who had been left out in the tundra too long with their crackpot theories. But Kirpotin did not fit that category. He had only recently been appointed vice-rector of his university. And the more I checked it, the more likely his story seemed. Larry Smith, of the University of California at Los Angeles, told me that the western Siberian peat bog was warming faster than almost any other place on the planet. Every year, he said, the spring melt was starting earlier and the rainfall was increasing, making the whole landscape wetter.

Others were finding big methane emissions in the region. Katey Walter, of the University of Alaska in Fairbanks, had told a meeting of the U.S. Arctic Research Consortium just a few weeks before about "hot spots" of methane releases from lakes in eastern Siberia that were "unlike anything that has been observed before." Peat on the bottom of lakes was converting to methane and bubbling to the surface so fast that it kept the lakes from freezing over in winter. And Euan Nisbet, of London's Royal Holloway College, who oversees a big international methane-monitoring program that includes Siberia, said his estimate was that methane releases from the western Siberian peat bog were up to 100,000 tons a day, which meant a warming effect on the planet greater than that of all the U.S.'s man-made emissions. "This huge methane flux depends on temperature," he said. "If peatlands become wetter with warming and permafrost degradation, methane release from peatlands to the atmosphere will dramatically increase."

So I wrote up Kirpotin's story for *New Scientist* magazine, emphasizing the methane angle. It went around the world. The London *Guardian* reproduced much of it the day after the story was released, under the front-page banner headline "Warming hits 'tipping point.'" In *Dr. Strangelove,* one nuclear device dropped on Siberia unleashed a thousand more. Here, in the real world of melting Arctic permafrost, one degree of global warming could unleash enough methane to raise temperatures several more degrees.

I had visited western Siberia a few years before, traveling with Western forest and oil-industry scientists to Noyabr'sk, a large oil town on the south side of the great peat bog. On a series of helicopter rides, I had seen thousands of square miles of still-intact swamp sitting on top of permafrost. The landscape was terribly scarred by human activity: divided into fragments by oil pipelines, roads, pylons, and seismic-survey routes; smeared with spilled oil; littered with abandoned drums, pipes, cables, and the remains of old gulags and half-built railways; and shrouded in black smoke from gas flares. The reindeer had fled, and the bears had been hunted almost to extinction. But the peat bog and the permafrost had survived. The helicopter landed frequently, and we jumped out without so much as getting our feet wet on the spongy surface.

No longer. On my way to meet Kirpotin's colleagues at their research station at Pangody, on the Arctic Circle, I flew for two hours over a vast bog that was seemingly going into solution. In place of the green carpet of moss and lichen, as Kirpotin had told me, numberless lakes stretched to the horizon. From the air, they did not look like lakes that form naturally in depressions in the landscape. They were generally circular, looking more like flooded potholes in a road. The lakes had formed individually from small breaches in the permafrost. Wherever ice turned to water, a small pond formed. Then surrounding lumps of frozen peat would slump into the water, and the pond would grow in an ever-widening circle, until mile after mile of frozen bog had melted into a mass of lakes.

"Western scientists cannot imagine the scale of the melting," Kirpotin told me. But I could see it beneath me as I flew east. It seemed to me that a positive feedback was at work, much as in the accelerated melting of Arctic sea ice. The new melted surface, darker than the old frozen surface, absorbed more heat and caused more warming. Kirpotin agreed. There seemed to be a "critical threshold" beyond which "the process of warming would be essentially and suddenly changed," he said. "Some kind of trigger hook mechanism would come into play, and the process of permafrost degradation would start to stimulate itself and to urge itself onwards." His imperfect English somehow made the events he was describing sound even more awful. "The problem concerned does not have only a scientific character: it has passed to the plane of world politics. If mankind does not want to face serious social and economic losses from global warming, it is necessary to take urgent measures. Obviously we have less and less time to act."

I was defeated in my efforts to see these processes in the Siberian bogs at first hand. Landing at Novy Urengoi with all the necessary paperwork, I was nonetheless refused admission. "You need a special invitation from an organization in the city," a fearsome policewoman at the airport told me as she confiscated my passport and put it in a safe. This was a company town. I later discovered that the mayor, a gas-company nominee, had won approval from Moscow some months earlier for special rules to keep out unwanted foreigners. Novy Urengoi was one of Russia's few surviving closed cities.

It was also rather disorganized. Unsupervised, I wandered into town anyhow, and looked around one of the most desolate and inhospitable places I have ever been to. No wonder its name means "godforsaken place" in the language of the local reindeer herders. I briefly met with the scientists I had come halfway across the world to see, before being rounded up and driven back to the airport by a spook wearing a double-breasted suit and a smile like Vladimir Putin's. He seemed to think I was a terrorist, and the fact that I was meeting scientists investigating the tundra only made him more suspicious. I can at any rate say I have been thrown out of Siberia.

Back home, concern has grown about the role of methane in stoking the fires of global warming. In early 2006, a dramatic study suggested that all plants, not just those in bogs, are manufacturing methane—something never previously considered by scientists. That led to headlines about trees causing global warming, which seemed a bit hard on them. If they do make methane, they also absorb carbon dioxide. And since trees have been around for millions of years, and there are probably fewer of them now than for the majority of that time, any role for them in recent warming seems unlikely. They are simply part of the natural flux of chemicals into and out of the atmosphere. Though if evidence emerged that they were emitting more methane than before, as a result of warming, that would be a big worry. And that is precisely what seems to be happening with peat bogs in the Arctic permafrost.

There is a critical line around the edge of the Arctic that marks the zone of maximum impact from global warming. It is a front line of climate change, marking the melting-point isotherm, where the average year-round temperature is 32°F, the melting point of ice. To the north of this line lie ice and snow, frozen soil and Arctic tundra. To the south lie rivers and lakes and fertile soils where trees grow. The line runs through the heart of Siberia and Alaska—where huge blocks of frozen soil, stable for thousands of years, are now melting—and across Canada, skirting the southern shore of Hudson Bay, through the southern tip of Greenland, and over northern Scandinavia.

Having failed to visit Kirpotin's field station to see the melting of the Siberian bogs close up, I went instead to northern Sweden to visit what is almost certainly the longest continually monitored Arctic peat bog in the world. In 1903, scientists took over buildings erected near Abisko during the construction of a railway to take iron ore from the Swedish mine of Kiruna to the Norwegian port of Narvik. They have been out there ever since, through the midnight sun and the long dark winters, measuring temperatures and dating when the ice came and went on Tornetrask, an adjacent lake; plotting movements of the tree line; examining the bog ecosystems; reconstructing past climates from the growth rings of logs in the lake, and investigating the cosmic forces behind the area's spectacular northern lights.

So they are on solid ground when they say it is getting dramatically warmer here. The lake freezes a month later than it did only a couple of decades ago—in January rather than mid-December. It used to stay frozen till late May, but several times in recent years an early breakup has forced the cancellation of the annual ice-fishing festival on the lake in early May. The average annual temperature here over the past century has been 30.7°F, but in recent years it has sometimes crept above 32°.

Just east of Abisko is the Stordalen mire. This is not a large bog, but it is old, and probably the best-monitored bog in the best-monitored Arctic region in the world. It has withstood numerous periods of natural climate change during the past 5,000 years. But suddenly it seems doomed. For here, as Kirpotin has found across the western Siberian wetland, the evidence of what happens to a bog that finds itself straddling the melting-point isotherm is obvious at every step. Apart from scientists, the bog's main visitors are birdwatchers. A couple of years ago, the local authorities built a network of duckboards for them. But already the boards are capsizing, because the mounds of permafrost on which they were built are melting and slumping into newly emerging ponds of water.

Arriving rather spectacularly aboard a helicopter hired to remove some equipment from the site, I found a dry hummock on which to talk to Torben Christensen, a Danish biochemist who heads the research effort here. "The bog is changing very fast," he said. Below our feet, the permafrost was still as deep as 30 feet, but a step away it was gone. We examined a crack in the peat, where another chunk was preparing to slide into the water. "Of all the places in the world, it is right here on the melting-point isotherm, on the edge of the permafrost, that you'd expect to see climate change in action," Christensen said. "And that is exactly what we are seeing. Of course, they are seeing it on a much bigger scale in the Siberian bogs, but here we are measuring everything."

Out on the mire, Christensen has some of the most sophisticated equipment in the world for measuring gas emissions in the air. In one area, individual bog plants grow inside transparent plastic boxes whose lids open and shut automatically as monitoring equipment captures and measures the flux of gases between plant and atmosphere. Pride of place goes to an eddy-correlation tower. This logs every tiny wind movement in the ambient air, vertical as well as horizontal, and uses a laser to measure passing molecules of methane and other gases. Combining the two sets of data, the tower can produce a constant and extremely accurate readout of the flux of methane coming off the mire.

There is a regular loss of methane from the bog now, says Christensen. Some of the gas seeps out of the boggy soil, some bubbles up through the pond water, and some is brought to the surface by plants. The figures sound small: an average of 0.0002 ounces of methane is released per 10 square feet of mire per hour. But scaled up, this packs a greenhouse punch. Combining the flux data with satellite images that show the changing vegetation on the Stordalen mire, Christensen estimates that in the past thirty years, methane emissions have risen by 30 percent and increased this small bog's contribution to global warming by 50 percent.

There is nothing unusual about Stordalen. It was not chosen to give dramatic results. Monitoring began back when researchers were intent only on tracking what they believed to be unchanging processes. Other local mires are faring far worse as the melting-point isotherm moves north. Out in the nearby birch forest, the Katterjokk bog has gone in just five years from being an area largely underlain by permafrost to being an ice-free zone. Rather, Stordalen looks to be typical of bogs across northern Scandinavia and right round the melting-point isotherm. Individually they are only a pinprick on climate change, but taken together they threaten an eruption.

Back in the warmth of the Abisko library, Christensen found a study showing that half the bog permafrost in the north of Finland has disappeared since 1975. The rest will be gone by 2030. Christensen himself has coordinated a study of methane emissions from peatlands at sites right around the Arctic, using temporarily deployed equipment for measuring gas fluxes. North of the melting-point isotherm, the study shows little change. Little methane bubbles out of the tundra in northeastern Greenland, for instance, where the average temperature is still around 14°F. But, he says, "as temperatures rise, methane emissions grow exponentially." The highest emissions are in western Siberia and Alaska, where big temperature rises are taking place.

What is happening out on these Arctic mires is, at one level, quite subtle. On many of them, temperatures remain cold enough to limit the decomposition of vegetable matter, and so carbon is still accumulating as it has done ever since the bogs began to form, at the end of the last ice age. But the decomposition rates are rising. And critically, because the melting permafrost is making the bogs ever wetter, more and more of the carbon is released not as carbon dioxide but as methane. That dramatically changes the climate effect of the bogs. Methane being such a powerful greenhouse gas, the warming influence of its release overwhelms the cooling influence of continued absorption of carbon dioxide. Thus "mires are generally still a sink for carbon, while at the same time being a cause of global warming," Christensen says. "This can be a hard point for people to grasp, but it is absolutely crucial for what is happening right around the Arctic."

There are still so few good data that it is hard to say for sure how much the Arctic peat bogs are contributing to global warming today. Current emissions of methane are probably still below 50 million tons a year. But that is still the warming equivalent of more than a billion tons of carbon dioxide. And with lakes forming everywhere, and climate models predicting that 90 percent of the Arctic permafrost will have melted to a depth of at least three yards by 2012, there is "alarming potential for positive feedback to climate from methane," says Christensen.

Larry Smith, of UCLA, estimates that the northern peat bogs of Siberia, Canada, Scandinavia, and Alaska could contain 500 billion tons of carbon altogether, or one third of all the carbon in all the world's soils. If all that carbon were released as carbon dioxide, it would add something like 5°F to average temperatures around the world. But if most of it were released as methane instead, it could provide a much bigger short-term kick. How much bigger would depend on how fast the methane was released, because after a decade or so, methane decomposes to carbon dioxide. If the methane all came out at once, it could raise temperatures worldwide by tens of degrees. That may be an unlikely scenario. Even so, the odds must be that melting along the melting-point isotherm is destined to have a major impact on the twenty-first-century climate. From Stordalen to Pangody, these bogs are primed.

15

THE ACID BATH

What carbon dioxide does to the oceans

The oceans are the ultimate sink for most of the heat from the sun and also for most of the greenhouse gases we are pouring into the atmosphere. The atmosphere may be the place in which we live and breathe, but for long-term planetary systems it is just a holding bay. At any one time, there is fifty times as much carbon dioxide dissolved in ocean waters as there is in the atmosphere. Given time, the oceans can absorb most of what we can throw into the atmosphere. But time is what we do not have, and the oceans' patience with our activities may be limited.

Carbon dioxide moves constantly between the oceans' surface and the atmosphere, as the two environments share out the gas. And, because of ever-rising concentrations in the atmosphere, the oceans currently absorb in excess of 2 billion tons more a year than they release. Much of that surplus eventually finds its way to the ocean floor after being absorbed by growing marine organisms—a process often called the biological pump. Sometimes there are so many skeletons falling to the depths that biologists call it marine snow.

Though they are the ultimate sink for most carbon dioxide, the oceans do not simply absorb any spare carbon dioxide left in the atmosphere. The relationship is much more dynamic—and much less reliable. In the long run, carbon dioxide seems to seesaw between the oceans on the one hand and the atmosphere and land vegetation on the other. Plants on land generally prefer things warm. Certainly the carbon "stock" on land is greater during warm interglacial eras like our own, and less during ice ages. By contrast, ocean surfaces absorb more carbon dioxide when the waters are cold. This seems to be partly because the plankton that form the basis of life in the oceans prefer cold waters, and partly because when the land is cold and dry, dust storms transport large amounts of minerals that fertilize the oceans.

During the last ice age, some 220 billion tons of carbon moved from the land and atmosphere to the oceans. This process didn't cause the ice ages, but it was a very powerful positive feedback driving the cooling. And that is a worry. For if the ice-age pattern holds, future generations can expect the oceans' biological pump to decline as the world warms. The story of the oceans' exchanges of carbon dioxide with the atmosphere may turn out to be rather like that of the carbon sink on land. In the short term, the extra carbon dioxide in the air has fertilized the biological pump and encouraged greater uptake. But in the longer term, warmer oceans are likely to weaken the biological pump and release large amounts of carbon dioxide into the air.

Is something of the sort likely? Very much so, said Paul Falkowski, of Rutgers University, in New Jersey, in a long review of the carbon cycle in *Science*. "If our current understanding of the ocean carbon cycle is borne out, the sink strength of the ocean will weaken, leaving a larger fraction of anthropogenically produced carbon dioxide in the atmosphere." With tens of millions of tons of carbon moving back and forth between the atmosphere and the oceans each year, it would take only a small change to turn the oceans from a carbon sink into a potentially very large carbon source. This may already be happening. In 2003, the NASA scientist Watson Gregg published satellite measurements suggesting that the biological productivity of the oceans may have fallen by 6 percent since the 1980s. It could be part of a natural cycle, he said, but it could also be an early sign that the biological pump is slowing as ocean temperatures rise.

So far, since the beginning of the Industrial Revolution, the oceans have absorbed from the atmosphere something like 130 billion tons of carbon resulting from human activities. While much of it has fallen to the seabed, a considerable amount remains dissolved in ocean waters—with a singular and rather remarkable effect: it is making the oceans more acid.

The carbonic acid produced by dissolving carbon dioxide is corrosive and especially damaging to organisms that need calcium carbonate for their shells or skeletons. These include coral, sea urchins, starfish, many shellfish, and some plankton. Besides eating away at the organisms, the acid reduces the concentration of carbonate in the water, depriving them of the chemicals they need to grow.

Acidity, measured as the amount of hydrogen ions in the water, is already up by 30 percent. To put it another way, the pH has dropped by 0.1 points, from 8.2 to about 8.1. If the oceans continue to absorb large amounts of the atmosphere's excess carbon dioxide, acidification will have more than tripled by the second half of this century, badly damaging ocean ecosystems. The most vulnerable oceans are probably the remote waters of the Southern Ocean and the South Pacific. They are distant from land, and so are already short of carbonate—in particular a form known as aragonite, which seems to be the most critical.

"Corals could be rare on the tropical and sub-tropic reefs such as the Great Barrier Reef by 2050," warned a report from Britain's Royal Society. "This will have major ramifications for hundreds of thousands of other species that dwell in the reefs and the people that depend on them." Other species may suffocate or die for want of energy. High-energy marine creatures like squid need lots of oxygen, but the heavy concentrations of carbon dioxide will make it harder for them to extract oxygen from seawater.

"It is early days," says Carol Turley, of the Plymouth Marine Laboratory, a world authority in this suddenly uncovered field of research. "The experiments are really only getting under way." But one set of results is already in. James Orr, of the Laboratoire des Sciences du Climat et de l'Environnement, in France, put tiny sea snails called pteropods into an aquarium and exposed them to the kind of ocean chemistry expected later in this century. These creatures turn up all around the world and are vital to many ecosystems. They are the most abundant species in some waters around Antarctica, where a thousand individuals can live in 300 gallons of seawater. As well as being a major source of food for everything from fish to whales, pteropods are the biggest players in the biological pump there.

Orr found that within hours, the acid pitted the pteropods' shells. Within two days, the shells began to peel, exposing the soft flesh beneath. In the real world, predators would break through the weakened shells. "The snails would not survive," he concluded. The demise of the pteropods would cause a "major reduction in the biological pump," the Royal Society agreed. Within a few decades, it could leave the oceans more acid than at any time for 300 million years.

Whatever the outcome, we are seeing the start of an unexpected and frightening side effect of rising atmospheric carbon dioxide levels. Perhaps the nearest parallel to the current situation was 55 million years ago—the last time a major slug of carbon was released into the atmosphere over a short period . . .

16

THE WINDS OF CHANGE

Tsunamis, megafarts, and mountains of the deep

It was Earth's biggest fart ever. Fifty-five million years ago, more than a trillion tons of methane burst from the ocean, sending temperatures soaring by up to 18°F extinguishing two thirds of the species in the ocean depths, and causing a major evolutionary shock at the surface. The story, while from long ago, is a reminder that methane lurks in prodigious quantities in many parts of the planet—not just in frozen bogs—and that one day it could be liberated in catastrophic quantities.

The first whiff of this prehistoric megafart was unearthed in 1991, from a hole drilled about a mile into a submarine ridge just off Antarctica. Examining the different layers of the ancient sediment removed from the hole, the geologists James Kennett, of the University of California at Santa Barbara, and Lowell Stott, of the University of Southern California in Los Angeles, found evidence of a sudden mass extinction of organisms living on the sea floor 55 million years ago. They had apparently disappeared from the ocean within a few hundred years—perhaps less. Kennett and Stott soon discovered that other researchers had detected evidence of similar extinctions from the same era, in Caribbean and European marine sediments. This was clearly a global event—one of the largest extinctions in the history of the planet.

What happened? Looking at the chemistry of fossils in the drilled sediment, the two geologists found some intriguing clues. There was, for instance, a sudden change in the ratio of two oxygen isotopes, known as oxygen-18 and oxygen-16. The ratio in the natural environment is very sensitive to temperature, and this isotopic "signature" in sediments and ice cores is a widely used indicator of past temperatures. Kennett and Stott concluded that after rising gradually for several million years, ocean temperatures had soared much more dramatically about 55 million years ago. The change happened at the same time as the extinctions.

The sediments also revealed a second isotopic shift, this time between isotopes of carbon. Earth's organic matter suddenly contained a lot more carbon-12. From somewhere, trillions of tons of the stuff had been released into the environment. Clearly a greenhouse gas, either carbon dioxide or methane, had caused both changes. The problem was finding a likely source with sufficient capacity to do the job.

Jerry Dickens, a biochemist at James Cook University, in Townsville, Australia, set himself the task of working out where this carbon-12 might have come from. The first suggestion was carbon dioxide in volcanic eruptions, which are a rich source of carbon-12 in the modern atmosphere. But, says Dickens, that would have required volcanic eruptions at an annual rate a hundred times the average over the past billion years. Fossil fuels like coal, oil, and natural gas were possible sources. But they are mostly buried out of harm's way, sealed in rocks. Given that there were no creatures digging them up and burning them at the time, that, too, seemed implausible. The same was true for methane from swamps and wetlands like those found today in Borneo and Siberia. About three times as many of them existed then, but even so, they could not have delivered the amount of carbon-12 required. Only one last source—big enough and accessible enough to unleash a climatic eruption—was left. That, Dickens suggested, had to be the vast stores of methane that geologists have recently been discovering frozen in sediment beneath the oceans: methane clathrates.

Methane clathrates are an enigma. They have until recently escaped the attention of oil and gas prospectors, because they don't turn up in the kind of deep and confined geological formations where prospectors traditionally look for fossil fuels. Nor are they the product of current ecosystems, such as tropical and Arctic bogs. They are generally close to the surface of the ocean floor but frozen—confined not by physical barriers but by high pressures and low temperatures, in a lattice of ice crystals rather like a honeycomb. Scientists still debate exactly how and when they were formed, but they seem to arise when cold ocean water meets methane created by microbes living beneath the seabed. Seismic surveys have revealed these structures in the top few hundred yards of sediments beneath thousands of square miles of ocean. They exist unseen, usually just beyond the edge of continental shelves. Many of these frozen clathrate structures trap even larger stores of gaseous methane beneath, where heat from Earth's core keeps them from freezing.

Dickens estimates that between 1 and 10 trillion tons of methane is tied up today in or beneath clathrates. But its confinement may not be permanent. Release the pressure or raise the temperature, and the lattices will shatter, pouring methane up through the sediment into the ocean and finally into the atmosphere. It seems that some such event must have happened 55 million years ago. Moreover, if this was the source of the great release of carbon-12, it would also explain why the extinctions appeared to be most serious in the ocean depths, where extensive acidification would have been almost certain. "Right now, most everybody seems to accept that the release of methane clathrates is the only plausible explanation for what happened 55 million years ago," says Dickens.

His chronology goes like this. For several million years, the world was warming, probably because of extraterrestrial influences such as the sun. The warming gradually heated sediments on the seabed until the clathrates started to shatter and release methane. Perhaps it happened in stages, with warming releasing methane that caused further global warming that released more methane. But at any rate, over a few centuries, or at most a few thousand years, trillions of tons of methane were eventually released into the atmosphere—enough to cause the observed global shift in carbon isotopes and a large and long-lasting hike in temperatures.

"The world just went into chaos," as Dickens puts it. Life on Earth was transformed almost as much as by the asteroid hit 10 million years before that wiped out the dinosaurs. Once the methane releases had ended, the planet's ecosystems gradually absorbed the remainder of the great fart, the climate recovered its equilibrium, and the oceans settled down again. But the evolutionary consequences of that long-ago event have lasted to this day. By the time the climate had recovered, many land and ocean species had become extinct, while others evolved and flourished.

"At the same time as the great warming, there was a major evolution and dispersal of new kinds of mammals," says Chris Beard, a paleontologist at the Carnegie Museum of Natural History, in Pittsburgh. It was "the dawn of the age of mammals." Among those on the evolutionary move were all kinds of ungulates—including the ancestors of horses, zebras, rhinos, camels, and cattle—and primates. And among the new primates evolving in the balmy conditions were the omomyids, the ancestors of simians, who in turn spawned humans.

Could such a cataclysm happen again? Maybe in the twenty-first century? Certainly there is still enough methane buried beneath the oceans. But could current global warming provide the trigger for its release? Some say that is unlikely; modern seawater is still much colder than it was 55 million years ago. But Deborah Thomas, of the University of North Carolina, who has analyzed the event in detail, is not so sanguine. The oceans may still be cooler, but they are also warming faster than they were 55 million years ago. And the pace of change may be as dangerous as the extent. If so, she says, "the trigger on the clathrate gun will be a lot touchier than it was 55 million years ago."

Apparently seaworthy ships can disappear from the ocean without warning for many reasons. They can be hit by giant waves, upturned by submarines, punctured by icebergs, or dashed onto rocks in storms. Could huge slugs of methane bursting from the ocean depths be another cause? Some say so. Take the strange case of Alan Judd and Witch Hole.

Judd is a British marine geologist at the University of Newcastle with a long interest in methane clathrates. In the late 1990s, he persuaded a French oil company to fund work in the North Sea to map giant pockmarks on the seabed. Geologists regard these otherwise inexplicable pockmarks as the aftermath of past methane eruptions from clathrates deep in the sediment. One day, about 90 miles off Aberdeen, Judd's remote-controlled probe was exploring a particularly large crater, about a hundred yards across and known to mariners as Witch Hole, when it crashed into something. Something large, metal, and unidentified, which destroyed the probe.

In the summer of 2000, Judd returned to try to find out what his probe had struck. This time he had money from a television company and a tiny remote-controlled submarine equipped with a video camera. He found the culprit. It was the steel hull of an 80-foot trawler dating, judging by its design, from the early twentieth century. The ship sat upright on the seabed, in the middle of the crater, apparently unholed. "The boat didn't go in either end first; it went down flat," Judd said later. "It looks as though it was just swamped." The ship could have gone down in a storm, but "for the boat to have randomly landed within Witch Hole would be an amazing coincidence," he said. "It is tempting to suggest that it is evidence of a catastrophic gas escape."

Efforts to identify the ship and find contemporary reports of why it went to the bottom have so far yielded nothing. And funds for another survey have failed to materialize. But the story remains an intriguing mystery to set beside other stories of ships that apparently disappeared in calm waters. Some say methane emissions from the depths could explain the mysterious loss of ships in the area of the Atlantic known as the Bermuda Triangle, for instance. Certainly, methane clathrates have been found in the area. So, while there is much mythology and misinformation about the Triangle, it may contain some truth. "When the gas bubbles to the surface, it lowers the density of the water and therefore its buoyancy," says Judd. "Any ship caught above would sink as if it were in a lift shaft." Any people jumping overboard to save themselves would sink, too. No trace would remain—at the surface.

Meanwhile, pockmarks are turning up on the seabed almost everywhere that clathrates are found: from the tropics to the poles, from the Atlantic, the Pacific, and the Arctic to the Indian and Southern Oceans. Evidence of when methane was released from the ocean floor remains sketchy, but the signs are of major releases. At Blake Ridge, off the eastern U.S., marine geologists have found pockmarks 700 yards wide and up to 30 yards deep, like huge moon craters. And drilling studies suggest that the ridge may still have around 15 billion tons of frozen methane hidden beneath the craters, with at least as much again trapped as free gas in warmer sediments beneath the frozen zone. European researchers have found pockmarks just as big in the Barents Sea southeast of Svalbard. The widely quoted estimate that 1 to 10 trillion tons of methane is trapped down there remains a bit of a stab in the dark, but the scale sounds right.

The lattice structures that hold methane clathrates survive only at low temperatures and high pressures, so sightings are rare. Occasionally they survive briefly at the ocean surface. Fishing nets bring lumps to the surface from time to time. They fizz away on the ship's deck, releasing their methane. Alarmed fishermen usually throw them back fast. Researchers have found white clathrate chunks "the size of radishes" sitting in the mud on the bottom of the Barents Sea; sometimes they track small plumes of methane rising from the seabed to the surface. Russian researchers have reported clathrates bursting out of the Caspian Sea and igniting "like a huge blowtorch, producing flames that rise several hundred metres high." But these events are mild curiosities compared with the evidence being pieced together of major catastrophic events caused by methane releases from beneath the ocean—including events that occurred much more recently than 55 million years ago.

On the east coast of Scotland, cliff faces often show a mysterious layer of gray silt about 4 inches thick sandwiched between layers of peat. The silt seems unremarkable, except that it extends right up the coast for hundreds of miles, and is full of the remains of tiny marine organisms that are normally found only on the ocean floor. The silt was deposited about 8,000 years ago by a tsunami that surged across the North Sea after the collapse of an underwater cliff on the edge of the continental shelf west of Norway. This was a huge event. The 250-mile-long cliff slumped more than 1.5 miles vertically down the slope onto the floor of the deep ocean, taking with it a staggering 1 billion acre-feet of sediment. It spread across an area of seabed almost the size of Scotland.

The scars left by this huge submarine slide were first spotted in 1979 by Norman Cherkis, of the Naval Research Laboratory in Washington, D.C. Cherkis was using sonar equipment to scour the ocean bed for hiding places for military submarines. He assumed at first that the slide had been caused by an underwater earthquake, though there was little seismological evidence for this. That presumption was shaken by a Norwegian marine geologist, Juergen Mienert, of the University of Tromso, who saw that the area of seabed that had slumped, known as Storegga, also contained large numbers of pockmarks associated with bursts of clathrates.

Mienert suggested that the slide coincided with a rise of 11°F in ocean temperatures off Norway as currents carrying the warm tropical waters of the Gulf Stream became much stronger in the aftermath of the last ice age. The strong wash of warm water over a previously cold seabed would have been enough, he said, to melt clathrates. Since just 100 cubic feet of clathrate contains enough methane to produce 16,000 cubic feet of gas at normal atmospheric pressure, the releases would have had explosive force, stirring up the seabed sediments over a huge area, and creating more releases and a cataclysmic slide.

Mienert estimated that this undersea eruption released between 4 and 8 billion tons of methane—enough to heat the global atmosphere by several degrees. His theory gained dramatic support when analysis of Greenland ice cores showed a big rise in methane concentrations in the air at that time. Some argue that the methane surge came from tropical wetlands that grew as the world warmed and became wetter. Mienert disagrees, but the argument has yet to be resolved.

The tsunami certainly had a huge impact. A 40-foot wave crashed into the Norwegian coast and deposited silt 20 feet above the shoreline in Scotland. The Shetland Islands took the brunt, receiving at least two waves that left a slimy trace 65 feet above what was then sea level. In the hours after Storegga slipped, many Stone Age people must have died on the shores of Europe. And it wasn't an isolated event. There appear to have been several earlier slips at Storegga. The fear must be that it could happen again here. "There is still a lot of methane on the north side of the slide," Mienert says.

Since the discovery of the Storegga slip, the remains of a number of other, similar slips have been discovered in areas of the ocean known to harbor methane clathrates. They have turned up off British Columbia, off both the East and West Coasts of the U.S., and at the mouths of great rivers like the Amazon and the Congo, where huge offshore fans of sediment contain methane generated by rotting vegetation from the rainforests upstream. Exactly when these slips occurred is not yet certain, but Mienert believes that the thermal shock caused by Storegga may have had a domino effect, releasing other clathrates stocks already made vulnerable by the warming postglacial oceans.

Some researchers postulate a "clathrate gun" theory of climate change, in which, at the end of the ice ages and perhaps at other times, successive releases of methane instigated a worldwide warming. They see the catastrophic event 55 million years ago as just the biggest in a whole family of methane-related climate disasters.

When I met Juergen Mienert in his lab, on a hill overlooking a fjord on the edge of Tromso (suitably raised, we joked, in case of a tsunami), he was planning a major new European research project to find more remains of slides and clathrate blowouts. The Euromargins project, which he chairs, "will be targeting areas where there are both pockmarks, indicating past clathrate releases, and warm ocean currents, indicating a risk of destabilization," he said.

He is already on the trail of an ancient slip high in the Arctic, off the north coast of Spitzbergen. This area is currently warming fast and is bathed periodically in warm waters from the Gulf Stream that break through the Fram Strait into the Arctic. "Some of the world's richest methane deposits lie right below that current," he said. He showed me new survey images of the seabed there, taken on a cruise two months before, in an area known as Malene Bay. They reveal another huge event. "Look at this," he enthused. "Look at the height of the cliff that fell. It was 1,500 yards high.

The prognosis, Mienert says, is worrying. Current conditions are disturbingly similar to those in which the great methane releases of the past happened—fast-rising sea temperatures penetrating the sediment and defrosting the frozen methane. Global warming, he believes, "will cause more blowouts and more craters and more releases." The risk of a giant tsunami blasting into Europe, the most densely populated continent on Earth, at the same time that a huge outburst of methane pushes climate change into overdrive is disturbing, to say the least.

Some argue that such concerns are exaggerated. It would take decades or even centuries for a warming pulse from the ocean to penetrate sediment to the zones where methane clathrates generally cluster. But Mienert counters that clathrates are being found ever closer to the surface, particularly in the Arctic. In any event, there is a second and much faster route downward for the heat. The U.S. naval researcher Warren Wood has discovered that seabed sediments often contain cracks that extend into the frozen clathrate zone. Warm water takes no time to penetrate the cracks and can quickly unleash the methane. As Richard Alley said of the crevasses inside ice caps, "Cracks change everything."

Methane is only the third most important greenhouse gas, after water vapor and carbon dioxide. But, says Euan Nisbet, "arguably it is the most likely to cause catastrophic change." This is "because the amount needed to change climate is smaller than for carbon dioxide, and because the amount of the gas available, in soils and especially methane clathrates, is so large." Methane has clearly had catastrophic effects in the past. In the dangerous world of sudden and unstoppable climate change, methane is the gunslinger.

IV

REFLECTING ON WARMING

17

WHAT'S WATTS?

Planet Earth's energy imbalance

Jim Hansen knows about the atmosphere from top to bottom. He began his career as an atmospheric physicist, studying under James van Allen, after whom the Van Allen Belts of the upper atmosphere are named. He published papers on the Venusian atmosphere before he moved on to our own. So when Hansen stops talking about degrees of temperature and starts counting how many watts of energy reach Earth's atmosphere and how many leave it, I recognize that we are getting down to the nitty-gritty of what sets Earth's thermostat.

I know about watts. I have a 60-watt bulb in the lamp over my desk. At school almost forty years ago, my physics teacher had a stock line for any lesson on electricity. "It's the watts what kill," he said, meaning that they are what matters. When Hansen says the sunlight reaching the surface of Earth in recent centuries has been about 240 watts for every 10.8 square feet, I can visualize that. It is four 60-watt bulbs shining on a surface area the size of my desk. That figure ever changes only slightly, because the sun itself is largely unchanging. If the sun were to grow stronger, more radiation would reach Earth, and we would warm up. But only so much. A warmed surface also releases more energy, so eventually a new equilibrium would be reached. Similarly, as additional greenhouse gases trap more solar energy, Earth warms until a new equilibrium is reached, with as much energy leaving as arriving. Put another way, Earth's temperature is whatever is required to send back into space the same amount of energy that the planet absorbs.

So what is happening today? Thanks to our addition of greenhouse gases to the atmosphere, the planet is suffering what Hansen calls "a large and growing energy imbalance" that "has no known precedent." The planet is warming, but it has not yet reached a new equilibrium.

The net warming effect of man-made pollutants is about 1.8 watts per 10.8 square feet. Most of this goes into heating either the lower atmosphere or the oceans. Ocean surfaces and the atmosphere share heat fairly freely, constantly exchanging energy. Because the oceans have a greater heat capacity than the atmosphere, they take the lion's share of the extra energy. But there are time lags in this exchange system. It takes some time to heat the oceans to their full depth. The warming of recent decades has created a pulse of heat that so far has gone as deep as 2,500 feet into the oceans in some places. As this pulse progresses, the oceans are draining more heat out of the atmosphere than they will once they return to a long-term balance with the atmosphere. It is rather like using a central heating system to warm a house. We have to heat all the water in all the radiators before the full effect of heating air in the house is felt. Likewise, the full impact of global warming will be felt in Earth's atmosphere only after the oceans have been warmed.

The best guess is that about 1°F—representing about 0.8 watts per 10.8 square feet—is currently lopped off the temperature of the atmosphere by the task of warming the oceans. That is warming "in the pipeline," says Hansen. Whenever we manage to stabilize greenhouse gases in the atmosphere, there will still be that extra degree to come. Half of it, Hansen reckons, will happen within thirty to forty years of stabilization, and the rest over subsequent decades or perhaps centuries.

While most of the extra heat being trapped by greenhouse gases is currently going into heating the oceans and the atmosphere, there is a third outlet: the energy required to melt ice. At present, no more than 2 percent is involved in this task. But Hansen believes that percentage is likely to rise substantially. Recent surging glaciers and disintegrating ice shelves in Greenland and Antarctica suggest that it may already be increasing. Melting could in future become "explosively rapid," Hansen says, especially as icebergs begin to crash into the oceans in ever-greater numbers.

There would be a short-term trade-off. Extra energy going into melting would raise sea levels faster but leave less energy for raising temperatures. But in the longer term, that would be of no help. For as more ice melts, it will expose ocean water, tundra, or forest. Those darker surfaces will be able to absorb more solar energy than the ice they replace. So we may get accelerated melting *and* more warming.

The critical term here is "albedo," the measure of the reflectivity of the planet's surface. Anything that changes Earth's albedo—whether melting ice or more clouds or pollution itself—will affect Earth's ability to hold on to solar energy just as surely as will changes in greenhouse gases. On average, the planet's albedo is 30 percent—which means that 30 percent of the sunlight reaching the surface is reflected back into space, and 70 percent is absorbed. But that is just an average. In the Arctic, the albedo can rise above 90 percent, while over cloudless oceans, it can be less than 20 percent.

During the last ice age, when ice sheets covered a third of the Northern Hemisphere, the vast expanses of white were enough to increase the planet's albedo from 30 to 33 percent. And that was enough to reduce solar heating of Earth's surface by an average of 4 watts per 10.8 square feet. It was responsible for two thirds of the cooling that created the glaciation itself. And just as more ice raised Earth's albedo and cooled the planet back then, so less ice will lower its albedo and warm the planet today.

According to the albedo expert Veerabhadran Ramanathan, of the Scripps Institution of Oceanography, if the planet's albedo dropped by just a tenth from today's level, to 27 percent, the effect would be comparable to a fivefold increase in atmospheric concentrations of carbon dioxide." To underline the importance of the issue, Ramanathan is organizing a Global Albedo Project to probe the albedo of the planet's clouds and aerosols. Lightweight robotic aircraft began flying from the Maldives, in the Indian Ocean, in early 2006. The project could prove as important as Charles Keeling's measurements of carbon dioxide in the air.

The prognosis for albedo cannot be good. We have already seen how the exposure of oceans in the Arctic is triggering runaway local warming and ice loss that can only amplify global warming. The same is also happening on land. Right around the Arctic, spring is coming earlier. And such is the power of the warming feedbacks that it is coming with ever-greater speed. As lakes crack open, rivers reawaken, and the ice and snow disappear, the landscape is suddenly able to trap heat. The "cold trap" of reflective white ice is sprung, and temperatures can rise by 18°F in a single day. No sooner have the snowsuits come off than travelers are sweltering in shirtsleeves.

Stuart Chapin, of the Institute of Arctic Biology, in Fairbanks, says that the extra ice-free days of a typical Alaskan summer have so far been enough to add 3 watts per 10.8 square feet to the average annual warming there. As a result, he says, the Arctic is already absorbing three times as much extra heat as most of the rest of the planet. And there are other positive feedbacks at work in the Arctic tundra. In many places, trees and shrubs are advancing north, taking advantage of warmer air and less icy soils. Trees are darker than tundra plants. And because snow usually falls swiftly off their branches, they provide a dark surface to the sun earlier than does the treeless tundra. Chapin estimates that where trees replace tundra, they absorb and transfer to the atmosphere about an extra 5 watts per 10.8 square feet.

This creates a surprising problem for policymakers trying to combat climate change. Under the Kyoto Protocol, there are incentives for countries to plant trees to soak up carbon dioxide from the atmosphere. They can earn "carbon credits" equivalent to the carbon taken up as the trees grow, and use these credits to offset their emissions from power stations, car exhausts, and the like. The idea is to promote cost-effective ways to remove greenhouse gases from the atmosphere—the presumption being that that will cool the planet. But in Arctic regions, the effect will usually be the reverse, because although new trees will indeed absorb carbon dioxide, they will also warm the planet by absorbing more solar radiation than the tundra they replace.

Clearly there is a balance between cooling and warming. But Richard Betts, of Britain's Hadley Centre, says that in most places in the Arctic, the warming will win. In northern Canada, he estimates, the warming effect of a darker landscape will be more than twice the cooling effect from the absorption of carbon dioxide. And in the frozen wastes of eastern Siberia, where trees grow even more slowly, the warming effect will be five times as great. Every tree planted will hasten the spring, hasten the Arctic thaw, and hasten global warming.

18

CLOUDS FROM BOTH SIDES
Uncovering flaws in the climate models

The graph flashed up on the screen for only a few seconds, but it set alarm bells ringing. Had I read it right? The occasion was a workshop on climate change at the Hadley Centre for Climate Prediction, held in Exeter in mid-2004. The room was packed with climate modelers from around the world. Even they raised a collective eyebrow when the graph sank in. If carbon dioxide in the atmosphere doubled from its pre-industrial levels, the graph suggested, global warming would rise far above the widely accepted prediction of 2.7 to 8.1°F. The real warming could be 18°F or even higher. Surely some mistake? Too much wine at lunch? No. This was for real.

Till now, climate modelers have graphed the likely effect of doubling carbon dioxide levels using what is known in the trade as a bell graph: the best estimate—about 5°—falls in the middle, and probabilities fall symmetrically on either side. So the chance that the real warming will be 8.1°, for instance, is the same as that it will be 2.7°. But the graph of likely warming that James Murphy, of the Hadley Centre, was displaying on an overhead screen that morning looked very different. The middle point of the prediction was much the same as everybody else's. But rather than being bell-shaped, the graph was highly skewed, with a long "tail" at the top end of the temperature range. It showed a very real chance that warming from a doubling of carbon dioxide would reach 10, 14, 18, or even 21°F.

Carbon dioxide is widely expected to reach double its pre-industrial levels within a century if we carry on burning coal and oil in what economists call a business-as-usual scenario. But nobody has seriously tried to work out what 18 degrees of extra warming would mean for the planet or for human civilization. It would certainly be cataclysmic.

Let's be clear. Murphy was not making a firm prediction of climatic Armageddon. But neither was this a Hollywood movie. The high temperatures on the display, he said, "may not be the most likely, but they cannot be discounted." Nor was Murphy alone with his tail. The meeting also saw a projection by David Stainforth, of Oxford University, that suggested a plausible warming of 21°F. Six months later, this new generation of scarily skewed distributions started turning up in the scientific journals. Unless the editors take fright, these figures will probably become part of the official wisdom, incorporated into the next report of the IPCC.

So what is going on? For one thing, modelers have for the first time been systematically checking their models for the full range of uncertainty about the sensitivity of the climate system to feedbacks that might be triggered by greenhouse gases. Assessing those efforts for the IPCC was the main task of the Exeter meeting. And what has emerged very strongly is that clouds, which have always been seen as one of the weakest links in the models, are even more of a wild card than anyone had imagined. The old presumption that clouds will not change very much as the world warms is being turned on its head. There may be more clouds. Or fewer. And their climatic impact could alter. It is far from clear whether more clouds would damp down the greenhouse effect, as previously thought, or intensify it. Being mostly of an age to remember 1970s Joni Mitchell songs, the climate scientists in Exeter mused over coffee that they had "looked at clouds from both sides now." And they didn't like what they saw.

An assessment of the sensitivity of global temperatures to outside forcing —whether changes in sunlight or the addition of greenhouse gases—has been central to climate modeling ever since Svante Arrhenius began his calculations back in the 1890s. This assessment mostly revolves around disentangling the main feedbacks.

The three biggest feedbacks in the climate models are ice, water vapor, and clouds. We have already looked at the effect of melting ice on the planet's albedo. It explains why the Arctic is warming faster than elsewhere and giving an extra push to global warming. Water vapor, like carbon dioxide, is a potent greenhouse gas, without which our planet would freeze. The story of what will happen to water vapor is a little less clear-cut. A warmer world will certainly evaporate more water from soils and oceans, and this process is already increasing the amount of water vapor in the atmosphere, amplifying warming. In the standard climate models, extra water vapor in the air at least doubles the direct warming effect of carbon dioxide. But it's when we come to clouds that the calculations get sticky.

A lot of water vapor in the air eventually forms clouds. At first guess, you might say that clouds would have the opposite effect of water vapor, shading us from the sun's rays and keeping air temperatures down. They do that on a summer's day, of course. But at night they generally keep us warm, acting like a blanket that traps heat. Globally, these two effects—or, rather, their absence—are most pronounced in deserts. Where there are no clouds, the days are boiling, but the nights can get extremely cold, even in the tropics.

The temperature effects of clouds turn out also to depend on the nature of the clouds. Their height, depth, color, and density can be vital, because different clouds have different optical properties. The wispy cirrus clouds that form in the upper atmosphere heat the air beneath, because they are good at absorbing the sun's rays and re-radiating the heat downward, whereas the low, flat stratus clouds of a dreary summer's day are good at keeping the air below cool.

Researchers still know surprisingly little about how many and what sort of clouds are above our heads. For instance, it has only recently emerged that there may be many more cirrus clouds than anyone had thought. Many are almost invisible to the naked eye, but nonetheless seem to be highly effective at trapping heat. Some studies suggest that, taken globally, the cooling and warming effects of clouds currently largely cancel each other out, with perhaps a slight overall cooling effect. But nobody is sure. And even small changes in cloudiness could affect planetary albedo substantially. If a warmer world tipped clouds into causing greater warming, the effects could be considerable.

So what is the prognosis? Again, a first guess is that extra evaporation will make more clouds, because a lot of the water vapor will eventually turn into cloud droplets. But even that may not be so simple. Evaporation doesn't just lift water vapor into the air to create more clouds; it also burns off clouds, leaving behind blue skies. And greater evaporation can also make clouds form faster, so that they fill with moisture faster, make raindrops faster, and dissipate faster. So, in a greenhouse world, fluffy cumulus clouds that we are used to seeing scudding across the sky all day could instead boil up into dark cumulonimbus clouds and rain out, leaving behind more blue skies.

Bruce Wielicki has been trying to figure out the answer to such questions during more than twenty years of cloud-watching at NASA's Langley Research Center, in Hampton, Virginia. He says that satellite data suggest that clouds probably still have an overall cooling effect on the planet; but, especially in the tropics, there is a trend toward clearer skies. Since the mid-1980s, the great tropical convection processes that cause air to rise where the sun is at its fiercest have intensified. As a result, storm clouds are forming and growing more quickly in those areas. This may be increasing the intensity of hurricanes across the tropics. Less obvious is Wielicki's discovery that the storm clouds not only form more quickly but also rain out more quickly. That leaves the tropics drier and less cloudy as a whole.

Many researchers see the phenomenon as strong evidence of an unexpected positive feedback to global warming. But Wielicki is cautious about what is behind his discovery of clearer tropical skies. We need to know, because the tropics are where an estimated two thirds of the moisture in the atmosphere evaporates—an important element in the planet's thermostat. "Since clouds are thought to be the weakest link in predicting future climate change, these new results are unsettling—the models may be more uncertain than we had thought," says Wielicki. His own guess is that clouds may be two to four times more important in controlling global temperatures than previously thought.

And that takes us back to the graphs on display in Exeter, where Murphy and Stainforth reached much the same conclusion as Wielicki in their new modeling projections of possible future warming. To make his graph, Murphy took a standard climate model and tweaked it to reflect the new range of uncertainties for cloud cover, lifetime, and thickness. His model responded by delivering much higher probabilities of greater-than-expected warming. "Variations in cloud feedback played a major role in the predictions of higher temperatures," he said. Susan Solomon, who as chair of the IPCC's science working group will be the final arbiter of what goes into its 2007 assessment of climate change, concurs. The biggest difference between models that give high estimates of global warming and those that give lower ones, she says, is how they handle cloud feedbacks.

Who is right? Are fears about a strong positive feedback from clouds warranted or not? One way of finding out is to test how the different models reflect the real world today. The IPCC is currently using this approach more widely to help weed out poor models from its analysis. Murphy has no doubt about what the outcome will be. The models that predict low warming "have a lot of unrealistic representations of clouds," he says. "The weeding process suggests higher temperatures." That is not proof, but it is worrying.

Clouds are not the only thing changing the reflectivity of Earth's atmosphere. Planet Earth is becoming hazier; the wild blue yonder is not so blue. The problem is pollution spreading across the Northern Hemisphere and much of Asia, blotting out the sun. The issue is not just aesthetic. Nor is it just medical, though millions of people die from the toxic effects of this pollution every year. It is also climatic. While some parts of the world are seeing temperatures soar, some of the world's most densely populated countries have seen temperatures drop. Climatologists who have spent many years warning about global warming are reaching the conclusion that we may need to be at least as concerned about the effects of this localized cooling.

The pollutants of concern here are normally lumped together under the name aerosols, but they are of many types and come from many sources. The culprits include operators of power stations in Europe, farmers burning crop stubble in Africa and trees in the Amazon, steel manufacturers in India, and millions of women cooking dinner over millions of open cooking stoves across the tropics. Most of these activities produce greenhouse gases, but they also produce aerosols in the form of smoke, soot, dust, smuts of half-burned vegetation, and much tinier but highly reflective sulfate particles. Depending on their characteristics, these aerosols reflect or absorb solar radiation. In fact, most do both, in varying quantities. But with one important exception that we shall return to, the dominant effect is cooling. The result is that some parts of the planet, from central Europe to the plains of India and the Amazon jungle to eastern China, have missed out on global warming either permanently or at certain times of the year.

A global cooling to counteract global warming might seem a good idea. Sadly, things are not so simple. The competing forces are pulling the climate system in two different directions that may not so much counteract as inflame each other. Certainly they introduce a new element of uncertainty in atmospheric processes. But although many countries are trying to reduce their emissions of smog-making aerosols, for excellent public-health reasons, the cleanup will lift the "parasol of pollution" over those countries. The likely result will be a burst of warming that could happen within days of the pollution's clearing.

We can see evidence of this already in central Europe. Fifteen years ago, countries like Poland, Czechoslovakia, and East Germany reeked with the smell of burning fossil fuels from the old Soviet-style heavy industries. Chimneys belched, and smog was endemic. The region where the three countries met became known as the "black triangle." The pollution was having a local cooling effect more than twice as great as the warming effect of greenhouse gases. Since the fall of the Berlin Wall, the old polluting industries have mostly shut down, and the air has cleared. More sun penetrates the smog-filled landscape, and central Europe has warmed correspondingly. In the past fifteen years, temperatures there have risen at three times the global average rate.

This real-world experiment shows clearly the power of aerosols to cool Earth's surface. And it raises another question for the future: How much warming is being suppressed globally by aerosols? "We are dealing with a coiled spring, with temperatures being held back by aerosols," says Susan Solomon, chief scientist for the IPCC. "If you shut off aerosols today, temperatures would increase rapidly, but we don't yet know exactly how much, because we don't know how coiled the spring is." The best guess until recently was that aerosols were holding back a quarter of the warming, or about 0.36°F. In other words, a greenhouse warming of 1.4 degrees since pre-industrial times has been reduced by aerosols to a current warming of 1 degree. But critics say this calculation is little more than a guess, and the first efforts at a more direct measurement of radiation changes caused by aerosols suggest that the spring may be much more tightly coiled.

I was present at one of the first meetings where these ideas were discussed in detail. The occasion was a workshop of climate scientists held at Dahlem, a quiet suburb of Berlin, in 2003. The meeting was discussing "earth system analysis," and the man who brought the issue to the table was the distinguished Dutch atmospheric chemist Paul Crutzen, whose brilliant and creative mind first divined many of the secrets of chemical destruction of the ozone layer. Back in the 1980s, Crutzen had stumbled on the notion that during a nuclear war, so many fires would be burning that the smoke "would make it dark in the daytime" and "temperatures would crash." That insight has led to continued analysis of the role of everyday aerosols in climate and to his conclusion, argued in Dahlem, that aerosols could be disguising not a quarter but a half to three quarters of the present greenhouse effect. "They are giving us a false sense of security," he said. Past calculations of the cooling effect of aerosols, he said, had been inferred by comparing the warming predicted by climate models with actual warming. The aerosol cooling effect was reckoned as the warming that had "gone missing." But as the modeler Stephen Schwartz, of the Brookhaven National Laboratory, put it on another occasion, "that approach assumes that we know that the climate model is accurate, which of course is what needs to be tested."

After dinner in Dahlem—over a few Heinekens, as I remember—Peter Cox, a hard-thinking, hard-drinking climate modeler then at the Met Office in England, did some back-of-the-coaster calculations about what Crutzen's conjecture might mean for future climate. He became rather absorbed. A couple of bottles later, he had come to the conclusion that, if Crutzen was right, the true warming effect of doubling carbon dioxide could be more than twice as high as existing estimates, at 12 to 18°F. The following morning, his more sober colleagues registered agreement. I went home and wrote a story for *New Scientist,* quoting Cox's numbers and the workshop's conclusion that the findings had "dramatic consequences for estimates of future climate change." I was rather excited by it, but the story decidedly failed to interest the rest of the world.

Later Cox, his Hadley Centre colleague Chris Jones, and Meinrat Andreae, of the Max Planck Institute for Chemistry, in Mainz, tested the guesses in more detail, and reached the same conclusions that Cox had on his coaster. They did it by running climate models that assumed either a low greenhouse warming moderated by a small cooling from aerosols, or a bigger greenhouse warming held back by a bigger aerosol cooling. They reported in *Nature* that the "best fit" involved a warming from doubling greenhouse gases that, without the moderating effect of aerosols, would be "in excess of" 10.8 degrees and "may be as high as" 18 degrees.

"Such an enormous increase in temperatures would be greater than the temperature changes from the previous ice age to the present," wrote the three researchers. "It is so far outside the range covered by our experience and scientific understanding that we cannot with any confidence predict the consequences for the Earth."

Still the world didn't take much notice. I asked Andreae about this strange indifference. "It's always amazing," he e-mailed me, "how many people don't see how important this issue is for the future development of the climate system." The discussion at the Dahlem meeting had rather changed his worldview, he said. "Before the Dahlem meeting, I was becoming kind of climate complacent, in the sense that I was convinced of coming global warming, but felt that it was going to be a couple of degrees and we could deal with that. Also, I felt that the aerosols were doing us a favor in slowing and reducing warming. But after it, I came to realize that the aerosols brake will come off global warming, and also that the aerosol cooling introduces a great uncertainty about climate sensitivity. I'm now in a situation where, as a human being, I hope that I'm wrong as a scientist. If we are right with our current assessment, there are really dire times ahead."

Models are only models, of course. But whatever the precise scale of the current aerosol effect, it would be quite wrong to imagine that it can carry on protecting us from the worst as global warming gathers momentum. That is because aerosols and greenhouse gases have very different life spans in the atmosphere. Aerosols stay for only a few days before they are washed to the ground in rain. By contrast, carbon dioxide has a life span of a century or more. If, for the sake of argument, we stuck with current emission levels of both aerosols and carbon dioxide, the aerosol levels in the air would stay the same. There would be no accumulation and no increase in the cooling effect. But carbon dioxide levels would carry on rising and produce ever greater warming.

Probably. The trouble is that scientific knowledge is, if anything, even poorer about aerosols than it is about the effects of clouds. Says Stephen Schwartz: "There are many different kinds of aerosols, lots of interactions among them, and unknown issues of cloud microphysics—all of which need to be better understood. This is hard science which I am afraid nobody has come to grips with yet." There is no dispute that some aerosols, such as sulfate particles from coal-fired power stations, predominantly scatter sunlight and reflect it back into space. They increase albedo and cool the planet for sure. Others, though, have some scattering effect but also absorb solar radiation and then re-radiate it, warming the ambient atmosphere. And with them it is harder to be sure where the balance between the two effects lies.

Here the biggest concern is soot, the black carbon produced from the incomplete burning of coal, biomass, or diesel. Scientific understanding of the role of soot is, to be frank, all over the place, as a quick scan of the major scientific journals makes clear. In March 2000, a paper in *Science* said soot was "masking global warming"; eleven months later another, in its chief rival, *Nature,* said soot was "generating global warming." Ten months later, presentations at a big U.S. conference of the American Geophysical Union called it variously "a cooling agent" and "the biggest cause of global warming after carbon dioxide." These can't both be right.

The truth seems to be this. A cloud of soot—whether from a forest fire, a cooking stove, or an industrial boiler—shields Earth from the sun's rays, thus cooling the ground beneath. But it also absorbs some of that radiation and converts it to heat, which it radiates into the surrounding air. So soot cools the ground but warms the air. The ground doesn't move, but the air does. The cooling effect, though intense, is mostly located near the pollution source; while the warming effect, though less intense, extends much farther.

There is still great uncertainty about the precise role of soot in global climate. Jim Hansen suggests that it could be responsible for up to a quarter of warming over parts of the Northern Hemisphere. He believes that soot may be the third most important man-made contributor to the greenhouse effect, behind carbon dioxide and methane, and that controlling it offers one of the cheapest, most effective, and quickest ways of curbing global warming. Even so, in those parts of the world where it is produced in large quantities, it is undoubtedly cooling the land. Those parts of the world are mainly in Asia. And now there is a new concern. Could aerosol emissions in India and China turn off the Asian monsoon?

19

A BILLION FIRES

How brown haze could turn off the monsoon

I have been traveling to India for twenty years now—not regularly, but often enough to notice that every time I go, the air seems to be dirtier and more choked with black smoke and fumes. In the cities, much of this comes from the exhaust pipes of the millions of ill-maintained diesel-burning buses and two-stroke rickshaws that ply the gridlocked streets. The haze also contains natural sea salt and mineral dust, a fair amount of fly ash and sulfur dioxide from India's coal-burning power stations, and huge amounts of organic material and soot from the countryside. For in India's million villages, where most of its billion-strong population still live, the air is often scarcely better than it is in the cities, with smoke billowing from a hundred million cooking stoves, all burning wood, dried cow dung, and crop residues.

This smoke is becoming a major climatic phenomenon. It is merging into one giant cloud that climate researchers call India's "brown haze." Its heart is over the northern Indian plain, one of the world's most densely populated areas, which suffers near-constant smog during the winter months. This is a giant version of the old pea-soup smog that used to hit London in the days when the city was heated by coal fires. As I complete this chapter, Delhi's air is reportedly worse than ever, with thick smog preventing flights from its airport. But the haze spreads more widely, shrouding the whole of India and beyond.

The term "brown haze" was coined by scientists during the first investigation of the phenomenon. In 1999, some two hundred scientists taking part in the Indian Ocean Experiment (INDOEX) assembled in the Maldives for a three-month blitz of measuring the air over India and the Indian Ocean from aircraft and ships. The results were a surprise, even to those who had planned the project. Every winter, from November to April, a pall of smog more than a mile thick occupied a huge area south of the Himalayas, stretching from Nepal through India and Pakistan, out over the Arabian Sea and the Bay of Bengal, and even south of the equator as far as the Seychelles and the Chagos Islands. It covered 4 million square miles, an area seven times the size of India.

"To find thick brown smog 13,000 feet up in the Himalayas, and over the coral islands of the Maldives, was a shock," says Paul Crutzen, one of the masterminds of the project. Crutzen, who won a Nobel Prize for predicting a dramatic thinning of the ozone layer fifteen years before it happened, said the haze had a similar potential to cause "unpleasant environmental surprise" in India and beyond. The haze could, he said, have "very major consequences" for the atmosphere.

The INDOEX findings proved controversial in India, which felt singled out for criticism. Why pick on us? locals asked. Indian government scientists issued a detailed and largely spurious "rebuttal." The INDOEX scientists quickly switched to discussing the "Asian brown haze"—and quite rightly, for the haze is an Asia-wide phenomenon. But when I used that term at a meeting in India in mid-2005, I was quietly hissed. Even mentioning an Asian haze is considered politically incorrect today. Why single out Asia? people ask. In fact, antagonism has become so great that many Indian scientists now refuse to discuss the subject with foreigners like me, for fear of getting into political hot water.

India has been the focus of attention because its aerosol pollution is of genuinely global importance. Dorothy Koch, of Columbia University, estimates that a third of the soot that reaches the Arctic, sending pollution meters soaring from Mount Zeppelin, in Svalbard, to northern Canada, comes from South Asia. The soot is falling onto the snow and ice, making the white surface darker and so triggering melting. When her findings were published, in April 2005, one headline read: "Home fires in India help to melt the Arctic icecap half a world away." No wonder the Indians are twitchy. Suddenly a country with one of the lowest per capita emissions of greenhouse gases in the world was being fingered as a prime cause of climate change.

But, wary though they may be in public, India's scientists have been at work finding out where all the pollution comes from. At first, they assumed that most must be the product of India's fast-growing and undoubtedly polluting industries. But at the Indian Institute of Technology, in Mumbai, they mocked up rural kitchens to check emissions from cooking stoves of the kind found across the Indian countryside. They fueled the fires with wood, crop waste, and dried cattle manure; on the stoves, they boiled kettles and even cooked lunch. They concluded that smoke emissions from India's domestic cooking fires produce between 1 and 2 million tons of aerosols a year, including a quarter of a million tons of soot. That makes them responsible for some 40 percent of India's aerosol emissions.

Discussion about the climatic impact of the Asian brown haze has become a statistical minefield. The "headline figure," widely quoted, is that in winter the haze reduces the amount of solar radiation reaching the ground in India by an average of about 22 watts per 10.8 square feet. That is a reduction of about a tenth, and would be enough to cause massive cooling. The statistic is literally true, but only part of the story. For only about 7 watts of that radiation is lost entirely, "backscattered" into space. The other 15 watts is absorbed by the soot in the aerosols and re-radiated, heating the atmosphere. Thus, though the radiation budget is much altered, the cooling effect is much less than it might otherwise be. Even so, in winter it is sufficient both to counteract global warming and to cool the air across much of India by an average of about 0.9°F. In summer, when the pollution is rained out in the monsoon and the skies are clearer, temperatures have risen in recent decades by about the same amount, in line with the global average.

The consequences don't end there, says Veerabhadran Ramanathan, the Indian scientist who, with Crutzen, masterminded INDOEX from the Scripps Institution of Oceanography. In particular, the cooling impact of the haze over the Indian land surface delays the heating of the land that stimulates the monsoon winds. It thus threatens the lifeblood of India: the monsoon rains.

There seems to be some confusion among scientists about the Indian monsoon. Scientists investigating the brown haze all claim that the monsoon has weakened in recent decades, and they see this as a likely effect of the haze. But researchers investigating global warming are equally certain that it has increased in intensity. What undisputed evidence there is suggests that the monsoon rains have become more intense in the traditionally wetter south of India, where the haze is thinner, but have diminished in the north, where the haze is thickest. How those trends develop is obviously of vast importance for a country entirely dependent on just a hundred days of monsoon rains to water the crops that feed a billion people. A wider collapse of the monsoon in South Asia would be a global calamity of immense proportions. It could happen.

East Asia could be in the same boat—a situation that would threaten food production for the world's most populous nation, China. North of the Himalayas, there is a similar intense brown haze in winter, though it is composed less of the smoke from burning cow dung and more of the sulfur dioxide and other fumes from burning coal. And it is interrupting the sun's rays. When Yun Qian and Dale Kaiser, of the U.S. government's Northwest National Laboratory, in Richmond, Washington, studied the records of Chinese meteorological sunshine recorders over the past fifty years, they found a decline in sunshine since 1980 of 5 or 6 percent in the most polluted south and east of the country.

And this decline is lowering temperatures. While global warming is evident across much of China, daytime temperatures in the most polluted regions have fallen by about 1°F. That, in turn, is altering rainfall patterns. In the south of the country, the monsoon rains are becoming stronger, with flooding in the great southern river, the Yangtze; whereas farther north, in the catchment of the Yellow River, there is now less rainfall. Chinese records, which are among the most meticulous in the world, suggest that this shift is the biggest alteration in the country's rainfall patterns in a thousand years. To some extent, links between the rainfall trends and the increasing brown haze are conjecture. But when climate models are programmed to include a strong Asian brown haze, many of them produce extra rainfall in southern China, coupled with near-permanent droughts in the north. So if the models are right, while the haze lingers, these major calamities are set to continue.

Meinrat Andreae estimates that about 8 billion tons of biomass is burned in the tropics each year—approaching 1 ton for every inhabitant of Earth. All of it produces aerosols that billow into the air.

Asian countries, with their huge populations, have the worst smog. But parts of Africa and the Brazilian Amazon are also shrouded when farmers clear land for crops by burning grasslands and forests. Hundreds of thousands of fires burn across the Brazilian Amazon each year, covering the area with billowing dense smoke. During the weeks of burning, the amount of sunshine reaching the ground typically drops by 16 percent. In Zambia, studies have found a 22 percent drop in sunlight as the savannah is burned.

The changes are "causing all sorts of havoc with the atmospheric circulation," says Dale Kaiser, of the Northwest National Laboratory, who is the author of the Amazon study. Over the Amazon, he says, the smoke causes cooling and suppresses the formation of raindrops. That both reduces rainfall and keeps the aerosols in the air longer. Meanwhile, the buildup of water vapor results in the upper atmosphere's becoming wetter, according to Daniel Rosenfeld, of the Hebrew University, who flew research planes through the smoke over the Amazon. It eventually forms a few extremely intense thunderstorms, known in the trade as "hot towers," which cause hailstorms. Hail falls in the Amazon only when fires have been burning.

Some of these changes could have impacts far beyond the regions where the smoke forms. Condensation in Amazon hot towers releases very large amounts of heat into the upper atmosphere, influencing jet streams and other wind patterns across the tropics and beyond. And more water vapor may reach the stratosphere, where it could increase ozone destruction. Meanwhile, modeling studies supervised by Jim Hansen suggest that soot emissions over India and China may trigger drought in the African Sahel and even warming in western Canada—though exactly how remains unclear.

These impacts are, of course, only the predictions of climate models. It is hard to prove whether they reflect events in the real world. But the models are based on real physical processes in the atmosphere. So at the least, they suggest the potential for a worldwide climatic change from the effect of aerosol emissions in the tropics. Cooking stoves in India, it seems, could have global consequences.

20

HYDROXYL HOLIDAY

The day the planet's cleaner didn't show up for work

It could be the doomsday that creeps up on us unawares: the day the atmosphere's cleaning service fails to show up for work. For one of the most disturbing secrets of our planet's metabolism is that just one chemical is responsible for cleaning most of the pollution out of the atmosphere. If it took a day off, we would be in serious trouble, with smog spreading unchecked across the planet.

The chemical in question is called hydroxyl. Its molecules are made up of one atom of oxygen and one atom of hydrogen. They are created when ultraviolet radiation bombards common gases such as ozone and water vapor. But it is the most ephemeral of chemicals. Almost as soon as it is created, it reacts with some other molecule, mostly some polluting substance, and is gone again. It has an average lifetime of about a second. Because it comes and goes so fast, it is also rather rare, with an average concentration in the atmosphere of less than one part per trillion. You could pack every last molecule of the stuff into the Great Pyramid of Egypt and still have room for two more atmospheres' worth.

Yet it is crucial to life on Earth. For hydroxyl is, more or less literally, the atmosphere's detergent. It transforms all manner of gaseous pollutants so that they become soluble in water and wash away in the rain. The process is called oxidation. To take one example, hydroxyl converts sulfur dioxide, which would otherwise clog up the air for months, to acid rain, which soon falls to the ground. Much the same happens to carbon monoxide and methane (both of which are oxidized to carbon dioxide), nitrogen oxide, and many others. The one major pollutant it doesn't neutralize is carbon dioxide, which, partly as a result, has a much longer lifetime in the atmosphere than most other pollutants.

Concentrations of hydroxyl are generally much higher in the warm air over the tropics, where ultraviolet radiation is most intense, but are close to nonexistent in the Arctic, where, despite ozone holes, there is usually little ultraviolet around to make more hydroxyl. As a result, "toxic chemicals that might survive for only a few days in the tropics will last for a year or more in Arctic air," says Frank Wania, of the University of Toronto. That is one reason, he says, why pollutants like acid hazes and pesticides accumulate in the Arctic, poisoning polar bears and much else.

Hydroxyl has a hard life keeping up with our polluting gases, especially since it is destroyed in the process of oxidizing them. Fears that the atmosphere's janitor could be overworked and in trouble go back a few years. But because the chemical is so transient and rare, it is virtually impossible to measure hydroxyl concentrations directly. All the estimates are indirect, based on measuring chemicals with which it reacts. So when Joel Levine, a NASA chemist, suggested back in the 1980s that hydroxyl in the air could have declined by 25 percent over the previous thirty years, his argument didn't make much headway, because he couldn't prove it. There was no chance of his producing something definitive like the Keeling curve on carbon dioxide.

In 2001, a brief forecast in the IPCC report of a possible 20 percent decline in hydroxyl by 2100, because of excess demands placed on it by a rising tide of pollution, met much the same fate. So did a report the same year by Ronald Prinn, a leading atmospheric chemist from the Massachusetts Institute of Technology, of a possible decline in global hydroxyl levels during the 1990s.

But we should be concerned. Hydroxyl spends more energy oxidizing one chemical than any other. That chemical is carbon monoxide. Emitted mostly from forest fires, fossil fuel burning, and small domestic stoves, it has for many years been the Cinderella pollutant. Dangerous to humans in confined spaces, it has been largely ignored as an environmental pollutant threat. The biggest concern has been that it oxidizes to carbon dioxide. But its concentration in the air tripled worldwide during the twentieth century. That suggests a bottleneck that could be the prelude to a wider breakdown of the cleaning service.

In the absence of good data on hydroxyl and its works, probably the best hope of finding a problem ahead of time is through modeling. Sasha

Madronich, of the National Center for Atmospheric Research, in Boulder, Colorado is one of the few researchers who have attempted to model how hydroxyl might respond to changing pollution levels. He says that the atmospheric cleaning service could have a breaking point: "Under high pollution, the chemistry of the atmosphere becomes chaotic and extremely unpredictable. Beyond certain threshold values, hydroxyl can decrease catastrophically." Many urban areas, he says, "are already sufficiently polluted that hydroxyl levels are locally suppressed." This is partly because the sheer volume of pollution consumes all the available hydroxyl, but also because the smog itself prevents ultraviolet radiation from penetrating into the air to create more.

"The oxidation processes that should clean the air virtually shut down in smog-bound cities like Athens and Mexico City," he says. It takes a breath of fresh air from the countryside to revive them. "If, in future, large parts of the atmosphere are as polluted as these cities are today, then we could anticipate the collapse of hydroxyl on a global scale." With large areas of Asia becoming submerged beneath a cloud of brown haze every year, it may be that the atmosphere is approaching just such a crisis. Nobody knows.

But the doomsday scenario may require another element. If the cleanup chemical is under pressure from too much dirt, the worst thing to happen would be a decline in supply of the chemical. So the critical question may be: What might reduce the amount of hydroxyl produced by the atmosphere? Clearly smog is a problem, because it reduces ultraviolet radiation in the lower atmosphere. But a thicker ozone layer, nature's protective filter against ultraviolet, could have the same effect. And the world is currently working quite hard to repair the damaged ozone layer and make it thicker. Our efforts to solve one environmental problem could exacerbate another.

The worry is that over the past thirty years or so, we have been living on borrowed time with hydroxyl. Pollutants like CFCs have thinned the ozone layer, and so let more ultraviolet radiation into the lower atmosphere. And while that is bad for marine ecosystems, and probably causes more skin cancers, it has ensured a beefed-up supply of hydroxyl to cleanse the air of many other pollutants. Arguably, it has helped the planetary cleaning service keep on top of a rising tide of pollution. Over the next half century, we should succeed in healing the ozone layer once again. There are good ecological, human-health, and even climatic reasons for doing this. But it could have a downside for hydroxyl.

So here is the doomsday scenario. If we repair the ozone layer, we will reduce hydroxyl production to the levels of the mid-twentieth century. But we will be doing it at a time when the demands on hydroxyl's services are considerably higher than they were then. That could be the moment when Madronich's threshold is crossed, and oxidation processes in the atmosphere go into sharp decline. I have no data, no models, and no peer-reviewed papers to justify this scenario. It is just that: a scenario and not a prediction. But it is plausible speculation. It could conceivably happen.

V

ICE AGES AND SOLAR PULSES

2 1

GOLDILOCKS AND THE THREE PLANETS

Why Earth is "just right" for life

Our sun has an inner ring of planets, starting with Mercury and moving out to Venus, Earth, and Mars. Right from their birth 5 billion years ago as cosmic debris, these planets have been more than lumps of rock. For one thing, they are hot, with thin solid crusts hiding large molten cores. Turbulent chemistry in their depths releases gases through the crusts. Although Mercury was too small, and its gravity too weak to capture these gases, the other three have held on to at least some of them, creating atmospheres. These atmospheres contain greenhouse gases such as carbon dioxide, water vapor, and methane that trap solar heat and create climates.

The three atmospheres of the three planets were initially probably rather similar. But they have evolved in very different ways. Today, Venus has a thick atmosphere with enough greenhouse gases to hold temperatures at around 850°F. Mars appears once to have had a considerable atmosphere and a climate that supported rainfall. It may have had life, as well. But somewhere along the way, it lost much of its atmosphere and dried up, and any life is now presumed extinguished. The demise of the life-support system on Mars is a conundrum, because the planet has plenty of carbon at its surface. It was probably once floating in the form of carbon dioxide in the atmosphere, where it would have formed a blanket sufficiently warm for liquid water and for life. But most of that carbon has ended up in rocks.

Earth, by contrast, has a rich and chemically very active atmosphere, and a sufficiency of greenhouse gases to maintain equable temperatures and lots of liquid water—and it is very much alive. Some planetary scientists have dubbed Earth the "Goldilocks planet." When, in the children's story,

Goldilocks tasted porridge at the house of the three bears, she found one bowl (Venus) too hot, one (Mars) too cold, and one (Earth) just right. At first, this seems the purest chance. Earth must have been just the right distance from the sun. And yet, since in the early days the three planets had very similar atmospheres, the theory has developed that their different fates had as much to do with the fates of those atmospheres as with the planets' distance from the sun.

Earth's atmosphere has certainly endured, and has proved a congenial place for the development of myriad life forms. Things were often difficult in the early days, it is true. At various points, the planet seems to have been entirely covered by ice and snow, with life surviving only in warm crevasses beneath the frozen exterior. The fate of Mars threatened. "It was a close call," says Joe Kirschvink, of the California Institute of Technology, in Pasadena, who coined the term "Snowball Earth" to describe this condition, which last occurred some 600 million years ago. He believes that the planet escaped a fate similar to that of Mars only because of a buildup of carbon dioxide emitted from volcanoes beneath the ice: "If the Earth had been a bit further from the Sun, the temperature at the poles could have dropped enough to freeze the carbon dioxide, robbing us of this greenhouse escape from Snowball Earth."

Despite such difficulties, Earth came through, and for the past half-billion years at least, it has maintained a surprisingly constant temperature. Not, as we shall see, completely constant, but surprisingly so given the cosmic forces being played out around it. In particular there was the sun. It is the main source of most of the energy and warmth at Earth's surface, of course. By comparison, the contribution of the heat from Earth's core is minute. But the sun has changed a great deal over the lifetime of Earth. Back in the early days—for about the first billion years of Earth's existence—it was a weak beast. It emitted about a third less energy than it does today. Even 500 million years ago, it was as much as 10 percent weaker than it is today. Yet, with Snowball Earth a distant memory, the world then seems to have been warmer than it is now, and ice-free. This is because the atmosphere was rich in methane, carbon dioxide, and water vapor, all forming a thick blanket that kept the planet and its growing armies of primitive life warm. Volcanic activity was still strong, so new releases of carbon dioxide topped up any leakage from the atmosphere, keeping concentrations around twenty times higher than they are today.

But as the planet has aged, the emissions from volcanoes have lessened, and carbon dioxide has gradually started to disappear from the atmosphere. Its decline may at various times have threatened a return of Snowball Earth, and a Martian relapse into a cold, lifeless world. But it may ultimately have saved the planet from a fate similar to that of Venus. This raises an interesting question. Did this happy Goldilocks outcome occur entirely by chance? Or could the planet have developed some kind of crude thermostat? The surprising answer is that it seems to have done just that.

Carbon dioxide, then as now, was removed from Earth's atmosphere largely by being dissolved in rain to form dilute carbonic acid. That acid ate away at rocks on the ground, which were made primarily of calcium silicate, creating calcium carbonate, which ended up as sediment on the ocean floor. This process has a temperature control built in, because the amount of rain depends on the temperature. So erosion rates rise when it is warm, but faster erosion removes more carbon dioxide from the air and lowers temperatures again. If the thermostat overshoots, and temperatures get too cold, then the rate of weathering slows, and temperatures recover. This is a negative feedback operating through the carbon cycle. It won't save us today, because it takes millions of years to have a serious impact. But over geological timescales, it was probably rather good at moderating temperatures and keeping the planet's climate convenient for life.

Very convenient. Suspiciously so, thought the charismatic British chemist and maverick inventor Jim Lovelock, back in the 1980s. Lovelock wondered if life itself might be controlling this process; and soon afterward two of his acolytes, Tyler Volk and David Schwartzman, suggested that he was right by demonstrating that basalt rocks erode a thousand times faster in the presence of organisms such as bacteria. This introduces a new and extremely dynamic negative feedback. More bacteria will keep the planet cool. But if the air gets too cool, the planet becomes covered by ice, the bacteria die, the erosion slows, and the atmosphere warms again. This process is potentially an extremely powerful thermostat for planet Earth, and is one of the foundation stones of Lovelock's grand vision of Earth as a self-regulating system called Gaia. It may also explain why the carbon cycle feedback did not save Mars: perhaps, at some critical moment, the red planet did not have enough life to make it work properly.

Lovelock is a controversial character. Now in his eighties, he first devised his idea of Gaia while working for NASA and trying to think of ways to decide if other planets had life. He figured that the best way was to look for signs of gases that could be made or maintained in the air only by life forms. And he began to realize that life could evolve quite naturally in ways that would maintain an environment that suited it. He argues that since the early days, life on Earth has evolved sophisticated strategies for stabilizing climate over long timescales. For him, the temperature of life on Earth was "just right" because life made it so by taking control of key planetary life-support systems like the carbon cycle.

For many years, Lovelock was virtually cast out of the scientific community, and Gaia was often seen as quasi-religious mumbo-jumbo. Major journals like *Nature* and *Science* would not publish his work. He made his living as a freelance inventor of scientific devices. But his idea of Earth as, metaphorically at least, a single living organism has made him the spiritual father of a whole generation of Earth system scientists. Whether or not you buy the notion of a living Earth, his way of thinking about Earth as a single system with its own feedbacks has been extremely influential.

The thermostat, whether run by life or by geology, is pretty crude. For some 400 million years, planet Earth has been getting cooler. Some see this as a refutation of Gaian ideas. But others, like Greg Retallack, a soil scientist at the University of Oregon, argue that the cooling happened because life, or at any rate large parts of it, wanted it that way. Plants in particular, he says, like it cool. And plants have proved extremely efficient at capturing carbon dioxide and burying it permanently where it cannot return to the atmosphere. Some 7 trillion tons of old vegetable carbon has been stored for tens of millions of years in the form of fossil fuels beneath Earth's surface. In addition, probably as much methane is captured in frozen clathrates beneath the ocean bed. That is a lot of warming stored away, as we are currently in danger of discovering the hard way.

The cooling of Earth has been a long, slow, and fitful process. Around 55 million years ago, as we saw earlier, Earth experienced the "biggest fart in history," a vast surge of methane into the atmosphere from the under-sea clathrate store, which pushed air temperatures up by around 9°F. That was clearly no part of a Gaian grand plan. But Gaians would argue that life-mediated feedbacks resumed control. The methane eventually decayed to carbon dioxide, which was in turn absorbed back into the oceans. But even after normality had been resumed, levels of carbon dioxide in the atmosphere were still about five times as high as they are today—at around 2,000 parts per million. Within a million years or so, however, those concentrations began to fall sharply. (Sharply, that is, on geological timescales: the average pace of decline was less than one ten-thousandth of the rate of increase in recent decades.) By 40 million years ago, they had subsided to 700 ppm. And by around 24 million years ago, they were below 500 ppm, probably for the first time since the planet's earliest days.

It was around then that an ice sheet spread across Antarctica—the first permanent ice to form on the planet for hundreds of millions of years. And by about 3 million years ago, another surge of cooling had begun, resulting in ice sheets forming in the Northern Hemisphere, too. Explanations for this general cooling range from continental drift in the western Pacific to another turn of the Gaian thermostat. But we can leave that to one side. Because the ice ages themselves—the geologically brief but extremely vicious cold snaps within the general cooling trend—happened on timescales of much more interest in our current climatic predicament. Unraveling the causes of the ice ages may, many climate scientists believe, provide vital clues to our fate in the coming decades.

THE BIG FREEZE

How a wobble in our orbit triggered the ice ages

The discovery that the world had once been plunged into an ice age was one of the great scientific revelations of the nineteenth century. It was to the earth sciences what Charles Darwin's theories on evolution were to the life sciences. It changed everything. The story emerged gradually, but the first man to perceive the scale of the glaciation that had overtaken so much of the Northern Hemisphere was a Swiss naturalist called Louis Agassiz. While Agassiz was summering in the Alps in 1836, his host pointed out giant scratch marks on the mountainsides that showed, he said, how the glaciers must once have extended much farther down their valleys.

Agassiz pondered the significance of this. He realized that he had seen similar marks in the landscape in many parts of Europe that were distant from present-day glaciers. He heard similar reports of glacial scratch marks from across North America. And he read contemporary newspaper stories of perfectly preserved mammoths being dug from the snow in Siberia, their meat so fresh that it was fed to local dogs and scavenged by polar bears. The only explanation, he concluded, was that much of the Northern Hemisphere must once have been covered by ice, and that the event happened very suddenly, in a vast, icy apocalypse. "The land of Europe, previously covered with tropical vegetation and inhabited by herds of great elephants, enormous hippopotami and gigantic carnivores, was suddenly buried under a vast expanse of ice," he wrote. "The movement of a powerful creation was supplanted by the silence of death."

Agassiz's vision was like a creation myth in reverse. Advances in geology soon revealed that not one ice age but a whole series of glaciations had occurred, separated by warm periods like our own. But his picture has otherwise survived remarkably intact. Indeed, recent evidence has revived his original idea that the onset of the last ice age must have been rather fast, with temperatures crashing in a couple of hundred years at most, and very probably much less.

We now know that two main ice sheets formed. One stretched from the British Isles across the North Sea and Scandinavia, and then west through Russia and western Siberia, and north across the Barents Sea as far as Svalbard. A second, even larger sheet covered the whole of Canada and southern Alaska, with a spur extending over Greenland. A smaller sheet sat over Iceland, and the seas around were full of thick floating ice. Strangely, northern Alaska and eastern Siberia, though deep-frozen, were never iced over. But, combined with the older ice covering Antarctica, these ice sheets contained three times as much ice as is present on Earth today—enough to keep sea levels worldwide some 400 feet lower than they are now—and covered 30 percent of Earth's land surface. The ice sheets were high as well as broad, rising up to 2.4 miles above the land surface. They chilled the air above and acted as a barricade for the prevailing westerly winds, which were forced south, skirting the ice sheets. This perpetuated the ice sheets, since the winds would have been the likeliest source of warmth to melt them.

Temperatures fell by around 9°F as a global average, but were 36 degrees lower than they are today in parts of Greenland, and just 5.4 degrees lower in the western Pacific Ocean. The world beyond the ice sheets became dry and cold. Deserts covered the American Midwest, France, and the wide lands of Europe and Asia between Germany and the modern-day Gobi Desert, in Mongolia. Farther south, the Sahara Desert expanded, the Asian monsoon was largely extinguished, and the tropical rainforests of Africa and South America contracted to a few refuges surrounded by grasslands. At the low point, around 70,000 years ago, even the grasslands were largely extinguished, leaving huge expanses of desert, from which winds whipped up huge dust storms. Humans lived by hunting on the plains and hunkering down in the small areas where lush vegetation persisted despite the cold and arid conditions.

It was clear from the start that something drastic must have triggered all this. Astronomical forces were suggested early on—in particular, the idea that the gravitational pull of other planets in the solar system, such as Jupiter, could influence the steady changing of the seasons, and in that way cause glaciers and ice sheets to grow. Many scientists of the day played with this idea. But the first man to subject it to detailed analysis was the son of a Scottish crofter with virtually no formal learning, but a passion for self-education and an extraordinary streak of diligence. James Croll was a shy, large-framed man with big ambitions. He stumbled on the idea of an astronomical cause for the ice ages while reading in libraries; transfixed, he spent most of the 1860s and 1870s pursuing the idea. He took numerous jobs, from insurance salesman to school caretaker to carpenter, in order to finance his passion.

Astronomical forces, he discovered, have three principal effects on Earth, all of which slightly alter the distribution of the solar radiation that reaches it. The effects are greatest in polar regions, where they can alter the amount of sun by as much as 10 percent. First, they change the shape of Earth's annual orbit around the sun. The orbit is not circular but slightly elliptical, and the shape of this ellipse changes according to the gravitational pull on Earth of the other orbiting planets. This "eccentricity" in Earth's orbit has a cycle that repeats itself about every 100,000 years.

As well as orbiting the sun once every year, Earth spins, making one revolution every day. But the axis around which it spins is not quite at a right angle to the direction of its orbit around the sun. So looked at from space, Earth appears to be spinning on a slight tilt. The combination of the orbit around the sun and the tilt of Earth's axis is what gives us our seasons, because it means that at certain times of the year the Northern and Southern Hemispheres see more or less of the sun. But this situation is not static. Astronomical forces also gradually alter the tilt of the axis. This change in Earth's "inclination" causes a difference in the intensity of the seasons. It has a 41,000-year cycle.

Finally, there is a further wobble in the axis around which Earth rotates, called the precession. This is exactly like the wobble that affects a spinning top. It influences the time of year when the different hemispheres are farthest from or nearest to the sun. It is complicated by its relationship with the other two effects, but it repeats on a cycle of 19,000 to 23,000 years. Currently the Northern Hemisphere has its summer, and the Southern

Hemisphere has its winter, when Earth is farthest from the sun; 10,000 years ago, it was the other way around.

It turns out that the eccentricity of Earth's orbit around the sun drives the 100,000-year cycles into and out of ice ages. Meanwhile, the other two effects, especially the precession, seem to trigger the short warm episodes that punctuate each ice age.

Croll realized that, averaged over a year, these changes made little difference to the amount of solar radiation reaching Earth. The overall effect was probably less than 0.2 watts per 10.8 square feet. But the changes did alter where and when the sun hit. Croll calculated in great detail how these influences waxed and waned over tens of thousands of years. And he established, at any rate to his own satisfaction, that they coincided with what geologists were then discovering about the timing of Earth's progress into and out of ice ages.

Taken together, the changing orbital shape, planetary tilt, and rotational wobble alter the strength of seasonality, making summers and winters more or less intense. And it was this that triggered the growth of ice sheets on land in the Northern Hemisphere, he said. Ice sheets would grow when northern winters were coldest. That would be when Earth was farthest from the sun, and when changing tilt ensured that it received the least sunlight. Once ice sheets started to grow, they would reflect ever more sunlight back into space, intensifying the cooling. Croll realized, too, that there was much less room for ice sheets to spread in the Southern Hemisphere, because they were confined to the continent of Antarctica. So the Northern Hemisphere would dominate events, driving the overall heat budget of the planet. But, he suggested, other feedbacks, such as changes to winds and ocean currents, could help drive the world further into an ice age.

In fact it turned out that Croll was wrong in assuming that it was cold winters that were critical. Later research proved that cold summers gave the world a bigger kick into ice ages, by providing little chance for winter accumulations of snow to melt. Nonetheless, Croll's work was a breathtaking piece of sustained cogent analysis that opened up a new field —much as Arrhenius did later with his examination of the impact of changing carbon dioxide levels on climate.

Croll's theory won him a few medals. But, being of low birth and of a taciturn disposition, he never fitted into the scientific salons of the day. They quickly tired of him and his ideas. Croll spent the last decade of his working life as the resident surveyor and clerk at the Scottish Geological Survey, in Edinburgh. To the last, he had to do his research in his own time. By the end of the nineteenth century, Croll and his ideas were largely forgotten. Even Arrhenius, who might have been expected to understand the importance of his work, dismissed it as an unwelcome rival to his own ideas, though in fact it complemented them.

Today, the idea that astronomical forces influence the formation of ice sheets is back in vogue and probably here to stay. Proof of its worth finally came in the 1970s. The British geophysicist Nick Shackleton carried out painstaking isotopic analysis of sediments on the ocean floor and in the process finally dated the glacial cycles sufficiently accurately to make clear their association with astronomical events. But even as the textbooks have been rewritten, Croll has been largely lost from the story. The orbital changes that he analyzed so painstakingly are known universally as the Milankovitch wobbles, after Milutin Milankovitch, a balding, monocled mathematician from Serbia who revived and elaborated Croll's ideas in the early twentieth century.

While Croll and Milankovitch have established to most people's satisfaction that orbital changes are the pacemaker of the ice ages, they did not by any means clear up the processes involved. How did a small change in the distribution of solar heating get amplified into a global freeze on a scale probably not seen since Snowball Earth thawed 600 million years before? And why, among a series of different wobbles, was it just one, with a return period of 100,000 years, that had much the greatest impact on global climate? A wobble, moreover, with an apparently weaker effect than the others on solar radiation reaching Earth. It seems, in the words of Dan Schrag, a geochemist at Harvard University, that Earth's system contains "powerful embedded amplifiers that can make it highly sensitive to relatively small forcings." Or, as Richard Alley would put it, we have a drunk on our hands. Identifying those amplifiers is important, not least because it should help answer how Earth's climate system might respond to our interference in its actions today.

Croll believed strongly in the power of growing ice itself to amplify cooling, and there is plenty of evidence to support the strength of this ice-albedo feedback. Once snow began to accumulate in the Canadian highlands around Hudson Bay, the ice sheet tended to grow of its own accord by cooling the area around it. Jim Hansen calculates that at the height of the last glaciation, it reduced the amount of heat absorbed by the planet's surface by some 4 watts per 10.8 square feet. What has troubled researchers rather more is exactly what limited it. Why, after reaching their greatest extent about 21,000 years ago, did the ice sheets begin to retreat?

Given the power of the ice-albedo feedback, it is far from clear why the ice sheets did not continue to grow until they had covered the entire planet and created a comeback for Snowball Earth. Even a change in the wobble to end the change in seasonality that started the ice growth might not have been enough. And it certainly would not explain the extremely fast collapse of the ice sheets at the end of the last glaciation. They disappeared more than ten times as quickly as they had arrived. Some fast feedback must have taken hold. One suggestion is that the sheer size of the ice sheets shut down further growth and eventually caused their rapid destruction. The main theory is that ice sheets are vulnerable to attack by heat rising from the interior of the planet. Trapped beneath the ice, it would have become of increasing importance as the sheets grew. Eventually, the theory goes, some threshold was passed, and the ice sheets melted from their base, creating a giant, continent-wide version of one of Hansen's "slippery slopes," with great chunks of ice skating into the ocean.

The second feedback that converted a planetary wobble into an ice age was greenhouse gases. Anyone who doubts the role of carbon dioxide in climate change should look at the graphs of atmospheric temperatures and of carbon dioxide levels in ice cores taken from the Greenland and Antarctic ice sheets. They cover the past half-million years, a period that includes several glaciations. Throughout, the two graphs are in lockstep. As carbon dioxide levels fall, so do temperatures, and vice versa. That does not determine which leads, but it clearly shows that they are engaged in a very intimate dance, in which carbon dioxide must amplify temperature changes even where it does not initiate them.

As temperatures fell at the start of each glaciation, around 220 billion tons of carbon left the atmosphere, returning during the brief interglacial periods. Its disappearance was enough to directly reduce Earth's uptake of solar energy by about 2 watts per 10.8 square feet. But what triggered this big shift in the planet's carbon cycle, and where did the carbon go? It certainly did not end up in vegetation on land, since that was shrinking as the world cooled. The obvious answer is the oceans. There are today about 44 trillion tons of carbon dissolved in the oceans—fifty times as much as in the atmosphere. So a minor uptake of carbon by the oceans could have had a huge effect on the atmosphere.

How might this have happened? Physics will help. Colder water (as long as it has not frozen) dissolves carbon dioxide better than warmer water. But most researchers believe that there must be some more dynamic feedback involved. To take a cue from Gaia, life is the obvious force here. One idea is that the initial cooling made the oceans more biologically productive. Plankton, the meadows of the oceans, do like colder temperatures. That is why the Southern Ocean around Antarctica is today one of the most productive. As the plankton grew, they drew more carbon dioxide out of the atmosphere. This strengthening of the biological pump would probably have been encouraged by enhanced dust storms, created by stronger winds and spreading deserts, which would have distributed mineral dust across the oceans. Even today, iron and other minerals are the limiting factor on the fecundity of much of the ocean food chain.

There may have been other feedbacks at work to push the planet into ice ages and drag it back out again. Methane may have been important. Its atmospheric concentration is in lockstep with temperature apparently as fixedly as that of carbon dioxide. One likely explanation is that the arid ice ages dried up wetlands and reduced their emissions of methane. Likewise, a colder atmosphere would have contained less water vapor—which would also have amplified the cooling.

A final amplifier may have been the ocean circulation system, with its huge ability to move heat around the planet. There is good evidence that the circulation system slows down during ice ages, and may have shut down entirely at the coldest point in the last glaciation. This is the province of a legend in the climate debate, Wally Broecker, and we will return to it in the next chapter.

The study of the ice ages suggests that over the past couple of million years at least, the natural climate system has constantly returned to one of two conditions. One is glaciated; the other is interglacial. The former has an atmosphere containing around 440 billion tons of carbon dioxide; the latter has an atmosphere containing about 660 tons. The planet oscillates between the two states regularly, repeatedly, and rapidly. But it doesn't hang around in any in-between states.

The evidence, says Berrien Moore III, the director of the Institute for the Study of Earth, Oceans, and Space, at the University of New Hampshire, "suggests a tightly governed control system with firm stops." There must be negative feedbacks that push any small perturbation back to the previous position. But there must also be strong positive feedbacks. Once things go too far, and the system seems to cross a hidden threshold, those positive feedbacks kick it to the other stable state. Each time, the guiding feedback seems to have rapidly moved about 220 billion tons of carbon between the atmosphere and the ocean.

That appears to have been the story for about the past two million years —until now. For the first time in a very long time, the system is being pushed outside this range. In the past century or so, human activity has moved another 220 billion tons of carbon into the atmosphere, in addition to the high concentrations of the interglacial state. The atmosphere now contains twice as much carbon as it did during the last ice age, and a third more than in recent interglacial eras, including the most recent. And we are adding several billion tons more each year. This extra carbon in the atmosphere has not been part of recent natural cycles. It comes mainly from fossilized carbon, the remains of swamps and forests that grew tens of millions of years ago.

This addition of carbon to the atmosphere is perhaps the biggest reason why Earth-system scientists feel the need to talk about the Anthropocene era. We are in uncharted territory. And the big question is: How will the system respond to this vast injection? Where will the carbon end up? There seem to be three possibilities. First, as some optimists hope, the system may deploy negative feedbacks to suppress change. Perhaps an accelerating biological pump in the ocean might remove the carbon from the atmosphere. It is possible. But the oceans generally like it cold. And there is no sign of such negative feedbacks kicking in yet, nor any obvious reason why they might. If anything, the biological pump has slowed in recent years.

The second possibility is the one broadly embraced by most climate models and the scientific consensus of the IPCC. It is that the system will carry on operating normally, gradually accumulating the carbon and gradually raising temperatures. There will be no abrupt thresholds that launch the climate system into a new state. This is moderately comforting, and fits the standard computer models, but it is contrary to experience over the past two million years.

And that raises a third possibility. Many Earth-system scientists think that their climate-modeling colleagues have not yet got the measure of the system. They fear that we may be close to a threshold beyond which strong positive feedbacks take hold, as they do when Earth begins to move between glacial and interglacial eras. The feedbacks may flip the system into a new, as-yet-unknown state. Most likely it would be one with much higher atmospheric concentrations of carbon dioxide and methane—more like the early days on planet Earth. That state might mean an era of huge carbon releases from the soil, or massive methane farts from the ocean floor, or wholesale changes to the ocean circulation system, or the runaway melting of the ice caps. That is conjecture. We simply don't know. But hold on to your hat: we could be in for a bumpy ride.

23

THE OCEAN CONVEYOR

The real day after tomorrow

Wally Broecker is a maverick—a prodigious and fearless generator of ideas, and one of the most influential figures in climate science for half a century. Sometimes he can be more. Amid the admiration for his science, you hear some harsh words about him in the science community. A bully, some say, especially to young scientists; a man who will use his influence to suppress ideas with which he disagrees. For a man in his seventies, he certainly comes on strong and relishes conflict. Here are his unprompted, on-the-record remarks to me about one of the U.S.'s leading climate modelers, who incurred the wrath of some Republican senators: "I think the senators were well out of line, but if anyone deserves to get hit, it was him. The goddamn guy is a slick talker and superconfident. He won't listen to anyone else. I don't trust people like that. A lot of the data sets he uses are shitty, you know. They are just not up to what he is trying to do."

Broecker is not a man to cross lightly. And to be honest, I thought a bit before writing the above. Much as I like his vigor, I'd hate to be caught in his crosshairs. Some believe he has earned the right to sound off about young colleagues he thinks don't pass muster. Some worry that Broecker seems to save his invective for people who resemble him in his younger years. But he is a man in a hurry. When I met him late in 2005, at Columbia's Lamont-Doherty Earth Observatory, his distinguished friend and collaborator Gerard Bond, a man a decade younger than Broecker, had recently died.

Broecker is a geochemist with an unimpeachable track record in pioneering the use of isotopic analysis to plot ocean circulation. He has been writing and thinking for more than three decades about what he calls the ocean conveyor, which more traditional scientists call the meridional over-turning circulation or the thermohaline circulation. Whatever you call it, it is the granddaddy of all ocean currents, a thousand-year circulation with "a flow equal to that of a hundred Amazon rivers," as he puts it.

The conveyor begins with the strong northward flow of the Gulf Stream pouring warm, salty water from the South Atlantic across the tropics and into the far North Atlantic. In the North Atlantic, the water is cooled, par-ticularly in winter, by the bitter winds blowing off Canada and Greenland. This cooling increases the density of the water, a process amplified by the formation of ice, which takes only the freshwater and leaves behind in-creasingly saline and dense water. Eventually the dense water sinks to the bottom of the ocean, generally in two spots: one to the west of Greenland, in the Labrador Sea, and the other to the east, down Wadhams's vertical chimneys. From there the water begins a journey south along the bed of the far South Atlantic, where a tributary, formed from cold, saline water plunging to the ocean bed around Antarctica, joins up. The conveyor then heads east through the Indian and Pacific Oceans before resurfacing roughly a thousand years later in the South Atlantic and flowing north again as the Gulf Stream to the far North Atlantic—where it goes to the bottom once more.

The circulation has many roles: distributing warm water from the trop-ics to the polar regions, mixing the oceans, and aiding the exchange of car-bon dioxide between the atmosphere and the oceans. Along the way, it keeps Europe anomalously warm in winter. In Richard Alley's words, it "allows Europeans to grow roses farther north than Canadians meet polar bears." On the face of it, the circulation is self-sustaining. The operation of the chimneys draws Gulf Stream water north, which provides more water for the chimneys. But it is also temperamental, prone to switching on and off abruptly. That switch, says Broecker, is a vital component of the entire global climate system. Not everyone agrees on the nature of the switch and how much it matters, but he makes a persuasive case.

Broecker's picture of the ocean conveyor is disarmingly simple. Too sim-ple, some say. He admits it had its origins in a cartoon. Asked by *Natural History* magazine to produce a diagram to illustrate a complicated argument about ocean-water movement, he drew a map with a few arrows suggesting likely "rivers" of intense flow within the circulation. "They sent it to an artist; he drew something, and I made a couple of corrections. I didn't realize it was going to be that important, but it was a popular magazine, and suddenly the diagram became a kind of logo for climate change."

Broecker is quite candid about the crudeness of the cartoon. But while some scientists might have disowned it, he recognizes its power and has embraced it. Its origins lie in Broecker's pioneering work using chemical tracers to identify movements of water in the oceans. He noticed that water in the Pacific and Indian Oceans appeared to be a mixture of water that had plunged to the depths in the North Atlantic and lesser amounts of water that had done the same thing around Antarctica. He could also see that water that had reached the ocean floor in the North Atlantic was largely made up of water that, prior to that, had made its way north as the Gulf Stream. To some extent, he filled in the rest. "The conveyor is clearly real," he insists. "But of course it's not as highly organized as it appears in the diagram." It is more a trend than a current—"a combination of random motions." And yet his cartoon has proved to be one of the most important concepts to emerge from climate science in the past quarter century.

Broecker chose the term "conveyor" because, he says, "I think names are very powerful, and that was much better than the proper scientific term. Some scientists say it is stupid, but laypeople can imagine a conveyor belt much more easily." He certainly has a way with words. Broecker was the first scientist to use the term "global warming," in a paper in the 1970s.

I first discovered the conveyor back in the late 1980s, while researching a book on environmental change. I was fascinated by the simplicity of the idea; by the fact that the conveyor might have two natural states, on and off; and by the scary possibility that climate change might shut the conveyor down if the ocean off Greenland became so flooded with freshwater that the dynamics of dense saline water formation around the chimneys broke down. For me, that idea was the first real inkling that climate change might not be as it was in the mainstream models—that the greenhouse effect might unleash something altogether nastier.

Early on, Broecker was often ambivalent about the potential for truly disastrous events. But by 1995, he felt confident enough to title a lecture on the conveyor to a big science conference "Abrupt Climate Change: Is One Hiding in the Greenhouse?" In it he outlined how evidence from sea-floor and lake sediments, ice cores, coral, and glacier records "demonstrates unequivocally" that an on-off switch on the global conveyor operated at the beginning and the end of the last ice age. The suggestion was that the conveyor had shut down and single-handedly started the ice ages, lowering temperatures by "4 degrees C [7.2°F] or more . . . often within the life-span of a generation"—a claim he inflated soon afterward, in the pages of *Scientific American*, to "10 degrees C [18°F] over the course of as little as a decade."

Broecker's picture, then, is of a powerful but fickle ocean conveyor with an on-off switch functioning in the far North Atlantic. Switched on, it warms the world, especially the Northern Hemisphere, and is typical of periods between ice ages. Switched off, it cools the Northern Hemisphere, and is typical of glaciations. But the system flickers at other times, too, he says. It triggered warm episodes that punctuated the depths of the last ice age, and perhaps drove more recent events such as Europe's medieval warm period and the little ice age. Broecker accepts that the ultimate forcing for these dramatic changes may lie in a celestial event like the slow movements of the Milankovitch cycles. But when a threshold is crossed and sudden climate change occurs, it is the conveyor that throws the switch.

These claims remain extremely controversial. Most would accept that Broecker is right that the conveyor slowed during the ice ages and probably shut down at various points. But most researchers believe that it was a consequence, and not a cause, of the glaciation. The big forces behind the cooling were the shift of carbon dioxide into the oceans and the spread of ice. And how important the ocean conveyor was in those processes has yet to be demonstrated. While the conveyor may have intensified cooling in the North Atlantic region, where the Gulf Stream is an acknowledged important feature in keeping the region warm, it is far less clear whether its global effects are anything like as big as Broecker claims.

But Broecker has rarely been bogged down in detail. Two years after making his claims for the ocean conveyor and the ice ages—and just a week before the world met in Japan to agree to the Kyoto Protocol—he was warning that climate change could trigger a future shutdown of the conveyor. "There is surely a possibility that the ongoing buildup of greenhouse gases might trigger yet another of those ocean reorganizations," he said. If it did, "Dublin would acquire the climate of Spitzbergen in ten years or less...the consequences would be devastating." He called the conveyor the "Achilles heel of the climate system."

Broecker was also, I think, making a wider point. He wants to generate a change in the way we think about the planet. Climate systems work, he suggests, rather as Stephen Jay Gould said evolution worked: not gradually, through constant incremental change, but in sudden bursts. Gould's phrase "punctuated equilibrium" sounds right for Wally's world of climate, too. And his new paradigm also fits the science of chaos theory, in which his ocean conveyor is an "emergent property" in the wider Earth system.

But the crux of the public debate on Broecker's ocean conveyor remains a very simple question: Could global warming shut the conveyor down? Broecker seems rarely to have doubted it. And the claim has in recent years seemed almost to have a life of its own. This struck me most strongly at a conference on "dangerous" climate change held at the Hadley Centre for Climate Prediction, in Exeter in 2005. There I met Michael Schlesinger, of the University of Illinois at Urbana-Champaign. He is a sharp-suited guy sporting a pastiche of 1950s clothes and hairstyle. But if there were serious doubts in Exeter about whether his style sense would ever come back into fashion, there was no doubt that his ideas about climate change had found their moment.

For more than a decade, Schlesinger has been making Broecker's case that a shutdown of the ocean conveyor could be closer than mainstream climate modelers think. Some critics feel that he just doesn't know when to give up and move on. But he has stuck with it, criticizing the IPCC and its models for systematically eliminating a range of quite possible doomsday scenarios from consideration. "The trouble with trying to reach a consensus is that all the interesting ideas get eliminated," he said at the conference. Science by committee ends up throwing away the good stuff—like the idea of the conveyor's shutting down. But in Exeter, Schlesinger was back in vogue. He had been invited to present his model findings that a global warming of just 3.6°F would melt the Greenland ice sheet fast enough to swamp the ocean with freshwater and shut down the conveyor. The risk, he said, was "unacceptably large."

Although he had been saying much the same for a decade, he was now considered mainstream enough to be invited across the Atlantic to expound his ideas at a conference organized by the British government. And he was no longer alone. Later in the day, Peter Challenor, of the British National Oceanography Centre, in Southampton, said he had shortened his own odds about the likelihood of a conveyor shutdown from one in thirty to one in three. He guessed that a 3-degree warming of Greenland would do it. Given how fast Greenland is currently warming, that seems a near certainty.

But all this is models. What evidence is there on the ground for the state of the conveyor? The truth is that dangerous change is already afoot in the North Atlantic. And, whatever the skepticism about some of Broecker's grander claims, the conveyor may already be in deep trouble. Since the mid-1960s, says Ruth Curry, of the Woods Hole Oceanographic Institution, the waters of the far North Atlantic off Greenland—where Wadhams's chimneys deliver water to the ocean floor and maintain Broecker's conveyor—have become decidedly fresher.

In fact, much of the change happened back in the 1960s, when some 8 billion acre-feet of freshwater gushed out of the Arctic through the Fram Strait. Oceanographers called the event the Great Salinity Anomaly. To this day, nobody is quite sure why it happened. It could have been ice breaking off the great Greenland ice sheet, or sea ice caught up in unusual circulation patterns, or increased flow from the great Siberian rivers like the Ob and the Yenisey. Luckily, most of the freshwater rapidly headed south into the North Atlantic proper. Only 3 billion acre-feet remained. Curry's studies of the phenomenon, published in *Science* in June 2005, concluded that 7 billion acre-feet would have been enough to "substantially reduce" the conveyor, and double that "could essentially shut it down." So it was a close call.

With the region's water still substantially fresher than it was at the start of the 1960s, the conveyor remains on the critical list. Another single slug of freshwater anytime soon could be disastrous. In the coming decades, some combination of increased rainfall, increased runoff from the land surrounding the Arctic, and faster rates of ice melting could turn off the conveyor. And there would be no turning back, because models suggest that it would not easily switch back on. "A shift in the ocean conveyor, once initiated, is essentially irreversible over a time period of many decades to centuries," as Broecker's colleague Peter deMenocal puts it. "It would permanently alter the climatic norms for some of the most densely populated and highly developed regions of the world."

As I prepared to submit this book to the publisher, new research dramatically underlined the risks and fears for the conveyor. Harry Bryden, of the National Oceanography Centre, had strung measuring buoys in a line across the Atlantic, from the Canary Islands to the Bahamas, and found that the flow of water north from the Gulf Stream into the North Atlantic had faltered by 30 percent since the mid-1990s. Less warm water was going north at the surface, and less cold water was coming back south along the ocean floor. This weakening of two critical features of the conveyor was, so far as anyone knew, an unprecedented event.

Probing further, Bryden found that the "deep water" from the Labrador Sea west of Greenland still seemed to be flowing south. But the volume of deep water coming south from the Greenland Sea, the site of Wadhams's chimneys, had collapsed to half its former level. The implication was clear: the disappearing chimneys that Wadhams had watched with such despair were indeed hobbling the ocean circulation. Broecker seemed on the verge of being proved right that the ocean conveyor was at a threshold because of global warming.

None of this demonstrated that Broecker's bleaker predictions of what would happen if the conveyor shut down were about to come true—that "London would experience the winter cold that now grips Irkutsk in Siberia." Something more like the little ice ages was the worst that most climate modelers feared. But there did seem to be a real possibility that many of Broecker's ideas were about to be put rather dramatically to the test.

24

AN ARCTIC FLOWER

Clues to a climate switchback

It must have felt like the springtime of the world. Anybody living on Earth 13,000 years ago could only have felt elation. An ice age of some 80,000 years was coming to an end. Temperatures were rising; ice was melting; rivers were in flood; and permafrost was giving way to trees and meadows across Europe and North America. In the Atlantic Ocean, the Gulf Stream was pushing north again, bringing warm tropical water and re-establishing an ocean circulation system that had shut down entirely in the depths of the ice age. Westerly winds blowing across the ocean were picking up the heat and distributing it across Europe and deep into Asia.

Meanwhile, in the tropics, the deserts were in retreat, the rainforests were expanding again from their ice-age refuges, and the Asian monsoon was kicking back in. Most spectacularly, the Sahara was bursting with life, covered in vegetation and huge lakes. This was the dawn of the age of *Homo sapiens,* who had supplanted the last of the Neanderthals during the long glaciation. If there had been a Charles Keeling around, he would have measured rising atmospheric levels of carbon dioxide and methane that were amplifying the thaw. He might even have invented the term "global warming" to describe it.

Then the unthinkable happened: the whole thing went into reverse again. Almost overnight, the thaw halted and temperatures plunged. Temperatures became as cold as they had been in the depths of the ice age. The forests returning to northern climes were wiped out; the permafrost extended; and ice sheets and glaciers started to regain their former terrain.

The springtime seemed to be over almost before it began. But this reversal was not the first. The previous 5,000 years had been full of them.

Some 18,000 years before the present, there was still a full-on ice age. By 16,000 years ago, the world was warming strongly. But by 15,000 years ago, it was cold again, with ice sheets reforming. At 14,500 years ago, it became so warm that within 400 years the ice caps melted sufficiently to raise sea levels worldwide by 65 feet. The cold gained the upper hand once more, only to give way to the pronounced warming of 13,000 years ago, which crashed again 12,800 years ago.

Today we can see this extraordinary climatic history recorded in ice cores extracted from the ice of Greenland and Antarctica. Graphs of the temperatures back then look like seismic readings during a big earth-quake—or cardiac readouts during a heart attack. They show a climate system in a protracted series of spasms. Looking back, we recognize the death throes of the ice age. But that is with hindsight. At the time, there was little evidence that the climate system had any sense of direction at all. It lurched between its glacial and interglacial modes. The one thing it didn't do was settle for a happy medium.

The last great cold snap of the ice age, 12,800 years ago, is known today as the Younger Dryas era. The dryas is a white Arctic rose with a yellow center that suddenly reappeared in European sedimentary remains, indi-cating that the old cold reasserted itself. The era is called the Younger Dryas to distinguish it from the Older Dryas, the climate reversal of a thousand years earlier, and the Oldest Dryas, which came before that. The Younger Dryas, like the others, was swift and dramatic. Within about a generation, temperatures fell worldwide—perhaps by as little as 3 to 5°F in the tropics, but by an average of as much as 28 degrees farther north, and, according to ice cores analyzed by George Denton, of the University of Maine, by 54 degrees in winter at Scoresby Sound, in eastern Greenland.

Not only temperatures crashed. Records of Chinese dust and African lakes and tropical trade winds and South American river flows and New Zealand glaciers all reveal dramatic changes happening in step 12,800 years ago. The world was much drier, windier and dustier. But in the Southern Hemisphere, temperatures may have gone in the opposite direc-tion. Marine sediment cores show dramatic warming in the South Atlantic and the Indian Ocean—as do temperature records in most Antarctic ice cores.

The Younger Dryas freeze lasted for fifty or so generations: 1,300 years. One can imagine tribes of *Homo sapiens* desperately relearning the crafts that got their ancestors through the ice ages. But it may also have triggered innovation. Some believe that dry conditions in the Middle East at the time may have encouraged the first experiments with crop cultivation and the domestication of animals. And then the freeze ended, and temperatures returned to their former levels even faster than they had fallen. Analysts of the Greenland ice-core chronology say publicly that the warming must have happened within a decade. But that is the minimum time frame for the change of which they can be certain, given the resolution of the ice cores. Richard Alley, who was there handling the ice cores, says: "Most of that change looks like it happened in a single year. It could have been less, perhaps even a single season. It was a weird time indeed." Like *The Day After Tomorrow,* only in reverse.

All this is doubly strange, because the Younger Dryas cooling went against the grain of all the long-term trends for the planet. The orbital changes that had triggered the glaciation had faded by then; astronomical forces were pushing the planet toward the next interglacial era. Of course, the real work was being done by feedbacks like melting ice, the return of greenhouse gases like carbon dioxide and methane into the atmosphere, and the revival of the ocean conveyor. These feedbacks would have turned a smooth progression into a series of jumps. But they would not easily have altered the direction of change. So why the backward flip? What made climate plunge back into the icy abyss when all the forcings and all the feedbacks should have been kicking the world into warmer times?

Chaos theory may help here. Alley says that it is just when conditions are changing fastest that the chances for seemingly random, unexpected, and abrupt change are greatest. The system is stirred up and vulnerable. The drunk is on a rampage. And there is a reasonable chance that some of the abrupt changes will be in the opposite direction to that expected. This is what, in the clever subtitle to his 2001 report on abrupt climate change, Alley called "inevitable surprise." What is equally clear is that at the time, the entire planetary climate system had just two possible states: glacial and interglacial. It knew no third way. And so, during the several thousand years when it was on the cusp between the two, it flickered between them.

On the ground, one element was a sudden switch in Broecker's ocean conveyor. It would be going too far to say that the Younger Dryas proves that the global conveyor is the great climate switch that Broecker claims. But the event makes a compelling case that events in the far North Atlantic can, without help from astronomical or any other forces, sometimes have dramatic and long-lasting effects on global climate.

The unexpected switch of the ocean conveyor was almost certainly triggered by melting ice. In the final millennia of the ice age, as melting made fitful but sometimes dramatic progress, a very large amount of liquid water was produced. Often it did not pour directly into the oceans but formed giant lakes on the ice or on land around the edges. The largest known of these is called Lake Agassiz, after the discoverer of the ice ages. It stretched for more than 600 miles across a wide area of the American Midwest, from Saskatchewan to Ontario in Canada, and from the Dakotas to Minnesota in the U.S., generally moving with the advancing front of warming.

In the early stages of the deglaciation, the lake drained south, down the Mississippi River into the Gulf of Mexico. But about 12,800 years ago, it seems, something stopped this and forced the lake to drain east. Perhaps the route south was blocked by land gradually rising after the weight of the ice was removed. Perhaps the lake simply passed over a natural watershed as it moved north with the retreating face of the ice sheet. But at any rate, there was eventually a huge breakout of freshwater from the heart of North America into the basin now occupied by the Great Lakes, and on into the North Atlantic.

The vast inrush of cold freshwater would have drastically cooled and freshened the ocean. High salinity was critical for sustaining the newly revived, and perhaps still precarious, ocean conveyor. So a fresher ocean shut down the conveyor once more. The warm Gulf Stream was no longer drawn north. Temperatures crashed across the North Atlantic region, and probably particularly around Greenland. The entire global climate system would have been shaken, and may have lurched back from its interglacial to its glacial mode.

Little of this narrative is cut-and-dried. The evidence is patchy. Some doubt whether even a vast eruption of freshwater down the Saint Lawrence Seaway would have had much influence on ocean salinity on the other side of Greenland. And others, hard-line opponents of the Broecker hypothesis, wonder exactly how important the ocean conveyor is to global climate. Even Broecker admits that parts of the story are "a puzzle."

But new evidence is emerging all the time. One compelling rewrite of the Broecker narrative has come from John Chiang, of the University of California at Berkeley. His modeling studies of the North Atlantic suggest that the most critical event at the start of the Younger Dryas may have been not the shutdown of the ocean conveyor itself but the impact of the freshwater invasion on the formation of sea ice in the North Atlantic. He says that an invasion that diluted the flow of warm water from the Gulf Stream would have rapidly frozen the ocean surface. The freeze itself would have flipped a climate switch, preventing further deepwater formation, sealing out the Gulf Stream, and, through the ice-albedo feedback, dramatically chilling the entire region.

Broecker has adopted this idea as an elaboration of his conveyor scenario. Some others see it as a replacement or even a refutation. Alley says: "It looks like this is the real switch in the North Atlantic. In the winter, does the water sink before it freezes, or freeze before it sinks? Sink or freeze. There are only two possible answers. That's the switch." Fresher, colder water will freeze; warmer, more saline water will sink. If the water sinks, the conveyor remains in place and the Northern Hemisphere stays warm. If it freezes, the circulation halts and the westerly winds crossing the ocean toward Europe and Asia stop being warmed by the Gulf Stream and instead are chilled by thousands of miles of sea ice. "The difference between the two is the difference in places between temperatures at zero degrees Celsius [32°F] and at minus 30 degrees [-22°F]," says Alley.

And that switch flipped, Alley argues, at the start and the finish of the Younger Dryas. At the start, freshwater invaded the North Atlantic; the ocean froze, and within a decade "there were ice floes in the North Sea and permafrost in the Netherlands." The westerly winds would have picked up the cold of the Atlantic ice and blown it right across Europe and into Asia. They would have cooled the heart of the Eurasian landmass, preventing it from warming enough to generate the onshore winds that bring the monsoon rains to Asia. This revised narrative also explains the concurrent warming in the Southern Hemisphere. If the Gulf Stream was not flowing north, the heat that it once took across the equator stayed in the South Atlantic. So as the North of the planet froze, the South warmed. A freshwater release in northern Canada had become a global climatic cataclysm. One, moreover, that went against all the long-term trends of the time.

It took about 1,300 years before the North Atlantic water switched back to sinking rather than freezing in winter. There is no consensus on what finally flipped the switch. But when it happened, it was at least as fast as the original freeze. The North Atlantic no longer froze; instead, the water was salty and dense enough to sink. The ocean warmed; the winds warmed; temperatures were restored in a year; nature returned to reclaim the tundra; and deglaciation got back on track.

For some, this story is encouraging. If it takes huge volumes of cold water flowing out of a lake to switch off the ocean conveyor, they say, we should be safe. There are no unstable lakes around of the kind created by the melting of the ice sheets. In any case, the world is warmer today than it was even at the start of the Younger Dryas. It may be, says Alley, that the world climate system is much more stable in warm times than in cold times. But equally it may not. For one thing, the superwarm world we are creating may contain quite different perils. For another, even the old perils may not have been neutralized as much as optimists think.

There is a cautionary tale in what happened 8,200 years ago. Despite large amounts of warming after the demise of the Younger Dryas cold event, the ice had one last hurrah. Again there was a large intrusion of cold freshwater into the North Atlantic. Again there was a big freshwater release; again the ocean was covered by ice; and again there seems to have been a disruption to the global conveyor. This was a lesser event than the Younger Dryas—probably only regional in its impact on climate, and lasting for only about 350 years. But it was nonetheless one of the biggest climate shifts of the past 10,000 years. And perhaps most significant for us today, says Alley, it happened in a world markedly more like our own than that of the Younger Dryas. Temperatures were generally rather close to those of today, and the ice sheets were quite similar. The event suggests, if nothing else, that if sufficient freshwater were to invade the North Atlantic today, it could have a similar impact.

As we have seen, in recent decades large slugs of freshwater have poured into the far North Atlantic. They may have come close to triggering a shutdown of the ocean conveyor. This trend is unlikely to end. As the climate warms and the permafrost melts in Siberia, river flows from there into the Arctic Ocean are rising strongly. And there is always the prospect of future catastrophic melting of the Greenland ice sheet, where glaciers are accelerating and lakes are forming.

Gavin Schmidt, one of Hansen's climate modelers at the Goddard Institute for Space Studies, says that the event 8,200 years ago is a critical test for today's climate models. "If we are to make credible predictions about the risks we run today of catastrophic climate change, those models need to be able to reproduce what happened 8,200 years ago," he says. "If we could do that, it would be really good. It could tell us a lot about processes highly relevant for the climate of the twenty-first century."

25

THE PULSE

How the sun makes climate change

The Arctic pack ice extended so far south that Eskimo fishing boats landed on the northern coast of Scotland. They didn't meet much opposition, because the hungry Highlanders had abandoned their crofts after grain harvests had failed for seven straight years, and had gone raiding for food in the lowlands to the south. In the 1690s temperatures in Scotland were more than 3°F below normal; snow lay on the ground long into the summer. Those who stayed behind were reduced to eating nettles and making bread from tree bark. The political repercussions of this Scottish turmoil are still with us today. The king became so worried by fears of insurrection that he shipped off angry clansmen and their starving families to set up Presbyterian colonies in Catholic Northern Ireland. And eventually, after widespread famine in the 1690s brought despair about the future for the Scots as a nation, the clan chiefs forged a union with England.

This was the little ice age: a climatic affair that began early in the fourteenth century and flickered on and off before peaking in the late seventeenth century and finally releasing its grip some 150 years ago. Like a mild echo of the ice ages, it spread its icy fingers from the north across Europe, pushing Alpine glaciers down valleys, creating spectacular skating scenes for the Dutch painters Breugel and Van der Neer, and allowing Londoners to enjoy the frolics of regular frost fairs on the frozen River Thames. On one occasion, Henry VIII traveled by sleigh down the river to Greenwich, and on another an elephant was led across the ice near Blackfriars Bridge.

There were some warm periods amid the cold. In the 1420s, an armada of Chinese explorers is reputed to have sailed around the north coast of

Greenland, a journey that would be impossible even in today's reduced Arctic ice. Between about 1440 and 1540, England was mild enough for cherries to be cultivated in the northeastern Durham hills. Much of Europe was exceptionally warm in the 1730s. But at the height of the little ice age, the Baltic Sea froze over, and there was widespread famine across northern Europe. Some suggest that half the populations of Norway and Sweden perished. Iceland was cut off by sea ice for years on end, and its shoals of cod abandoned the seas nearby for warmer climes. Some say the cold was the hidden hand behind the famine, rising grain prices, and bread riots that triggered the French Revolution in 1789.

In North America, tribes banded together into the League of the Iroquois to share scarce food supplies. The Cree gave up farming corn and went back to hunting bison. But the era was symbolized most poignantly by the collapse of a Viking settlement founded in the balmy days of the eleventh century by Leif Erikson. The Viking king had a real-estate broker's flair for coining a good name: he called the place Greenland to attract settlers. The settlement on the southern tip of the Arctic island thrived for 400 years, but by the mid-fifteenth century, crops were failing and sea ice cut off any chance of food aid from Europe.

If the Viking settlers had followed the ways of their Eskimo neighbors and turned to hunting seals and polar bears, they might have survived. But instead, they stuck to their hens and sheep and grain crops, and built ever-bigger churches in the hope that God would save them. He did not. When relief finally arrived, nobody was left alive in the settlement. Creeping starvation had cut the average height of a Greenland Viking from a sturdy five feet nine inches to a stunted five feet. The last women were so deformed that they were probably incapable of bearing a new generation. We know all this because their buried corpses were preserved in the spreading permafrost.

The little ice age, first documented in the 1960s by the British climate historian Hubert Lamb, is now an established part of Europe's history. It has often been seen as just a historical curiosity—a nasty but local blip in a balmy world of European climatic certainty. But it is increasingly clear that what Europe termed the little ice age was close to a global climatic convulsion, which took different forms in different places.

Because it came and went over several centuries, the task of attribut-
ing different climate events around the world to the influence of the little
ice age is fraught with difficulties. But reasonable cases have been made
that it blanketed parts of Ethiopia with snow, destroyed crops and precip-
itated the collapse of the Ming dynasty in seventeenth-century China, and
spread ice across Lake Superior in North America. In the tropics, temperatures were probably largely unchanged, but rainfall patterns altered substantially. In the Amazon basin, the centuries of Europe's little ice age were
so dry that fires ravaged the tinderbox rainforests. In the Sahara, which often seems to experience climate trends opposite to those in the Amazon,
repeated floods in the early seventeenth century washed away the great
desert city of Timbuktu.

The little ice age is not the only climate anomaly in recorded history.
Another, known because of its influence on European climate as the medieval warm period, ran from perhaps 800 to 1300, ending just as the little ice age began. Because it is rather more distant than the little ice age,
its history and nature are rather less clear. Certainly, at various times grains
grew farther north in Norway than they do today, and vineyards flourished
on the Pennines, in England. Warmth brought Europe wealth. There was
an orgy of construction of magnificent Gothic cathedrals. The Vikings, as
we have seen, set up in Greenland at a time when parts of it could certainly
be described as green. Some claim that the medieval warm period may have
been warmer even than the early twenty-first century. But most researchers
are much more cautious.

Reconstructions of past temperatures come mainly from looking at the
growth rings of old trees. There are exceptions, but generally, the wider
the rings, the stronger the annual growth and the warmer the summer.
Keith Briffa, a British specialist in extracting climate information from
tree rings, says: "The seventeenth century was undoubtedly cold. The evidence that the period 1570 to 1850 was also cold seems pretty robust. But
the medieval warm period is still massively uncertain. There is not much
data, and so much spatial bias in the data. We think there was a warm period around AD 900, certainly at high northern latitudes in summer, where
we have the tree-ring evidence. But we have virtually nothing else." It
looks likely that much of Europe was between 1.8 and 3.6°F warmer in the
medieval warm period than it was in the early twentieth century, while the

little ice age was a similar amount cooler in Europe. But any global trends were almost certainly much smaller.

In any case, to talk about a medieval warm period at all is, in the view of many, a very Eurocentric view. Tree rings from the Southern Hemisphere show no sign of anything similar there. Indeed, away from the North Atlantic, those centuries were, if anything, characterized by long superdroughts that caused the collapse of several major civilizations. In Central America, the Mayans had thrived for 2,000 years and built one of the world's most advanced and long-lasting civilizations. Theirs was a sophisticated, urbanized, and scientific and technologically advanced society of around 10 million people, with prolific artistic activities and strong trade links with its neighbors, and seemingly every resource necessary to carry on thriving—strikingly like our own in many respects. Yet faced with three decades-long droughts between the years 800 and 950, which may have been the worst in the region since the end of the ice age, the entire society crumbled, leaving its remains in the jungle. A few hundred miles north, a number of advanced native North American societies collapsed under the impact of sustained droughts through the American West. Best documented are the Anasazi people, ancestors of the modern Pueblo Indians. They had built elaborate apartment complexes in the canyons of New Mexico, and had developed sophisticated irrigation systems for growing crops, but were forced to flee into the wilderness after a long drought that peaked in the 1280s.

The little ice age and the medieval warm period appear to have been recent natural examples of climate change. Though the warming and cooling implied in their names may have been restricted largely to the North Atlantic region, they seem to have left a signature in glaciers and megadroughts across the planet. So what caused them? And does it have anything to tell us about our own future climate? Many theories have been advanced.

The pendulum moves too fast for any orbital cycles. Some theorists have suggested a role for volcanic eruptions, which shroud the planet with aerosols that can cool it. It is true that at certain times during the little ice age, there were major eruptions. The year after the eruption of Tambora, in Indonesia, in 1815, crops failed from India to Europe and North America. It became known as "the year without a summer." But volcanic dust clouds cool temperatures for only a few years at most. They may from time to time have exacerbated the cooling, but they were not sufficiently frequent or unusual to explain a cold era that lasted on and off for almost half a millennium.

Most climatologists believe that the sun should get the blame. The coldest part of the little ice age, in the mid-to-late seventeenth century, is known as the Maunder Minimum. The popularizing of the telescope by Galileo a few decades before meant that astronomers of the day were able to note the virtual disappearance between 1645 and 1715 of the by-then-familiar spots on the surface of the sun. This is now recognized as a good indicator of a reduced output of solar energy. The best guess is that solar radiation reaching Earth's surface during the Maunder Minimum fell by perhaps half a watt per 10.8 square feet, or around 0.2 percent. But climatologists find it perplexing that such a widespread effect could result from such a modest change.

Enter an idiosyncratic, larger-than-life researcher working at the Lamont-Doherty Earth Observatory, just down the corridor from Wally Broecker. His name was Bond, Gerard Bond. Like Broecker, he hated getting bogged down in detail, and liked seeing the big picture. Like Broecker, he was willing to fly a kite, trusted his intuition, and had the confidence to propose an idea in public just to see if anyone could shoot it down. And, again like his compatriot, he had the intellectual reputation to get his kite-flying published in the often conservative scientific literature.

Bond argued forcefully until his death, in 2005, that the little ice age and the medieval warm period were the most recent signs of a pervasive pulse in the world's climatic system. This pulse, he said, had a cycle that recurred once every 1,500 years or so. It was a pulse, moreover, that seemed largely unaffected by other, apparently bigger influences on global climate, like the Milankovitch orbital cycles that triggered the major glaciations. Ice age or no ice age, he argued, the pulse just kept on going. Bond didn't invent the pulse out of thin air. Other researchers had unwittingly been on its trail for years. But, like his friend down the corridor, Bond was the man who had the confidence to compose a big picture out of the scattered fragments of evidence.

In the early 1980s, a graduate student in Germany made the first breakthrough. While at the University of Göttingen, Hartmut Heinrich was examining cores of sediment drilled from the bed of the North Atlantic. He found a number of curious layers of rock fragments that showed up in cores drilled as far apart as the east coast of Canada, the waters west of the British Isles, and around Bermuda. Radiocarbon dating revealed that these rock fragments were laid down in at least six bands over the 60,000 years before the end of the last glaciation, at intervals of roughly 8,000 years.

I looked at some of these rock fragments in the marine sediment store at Bond's old laboratory in New York. They are enormously distinctive. A browse among the trays of sediment revealed fairly subtle differences among the different cores: a change of color here, a slightly different consistency of dust there. Almost everything in these sediments has gone through the mill of being eroded from Earth's surface, discharged down rivers, and dumped in tiny bits on the seabed. But then there are Heinrich's layers. These are a mass of stones the size of gravel or pebbles, but sharp-edged and clearly untouched by the normal processes of erosion and deposition. Researchers soon gave the events that produced them their own name: Heinrich events. There was nothing like them in the sediment record.

Apart from their size and shape, something else was odd about these rock fragments. Though they had been found way out in the middle of the Atlantic Ocean, geologists swiftly established that they came from the Hudson Bay area of northern Canada. How could they have got so far offshore and so far south? What took them there? The only logical explanation, given that all the Heinrich events took place during the last glaciation, was that they had been ripped from the bedrock by great glaciers and carried south on the underside of icebergs. They traveled a long way because the North Atlantic was extremely cold, and were eventually dumped onto the ocean floor as the icebergs melted. That raised other questions. What climatic events would send vast armadas of icebergs sailing south into the tropics? And why the apparent 8,000-year cycle?

The next clue came a few years later, in the early 1990s, when a distinguished Danish glaciologist, Willi Dansgaard, of the University of Copenhagen, discovered in the Greenland ice-core record a series of large and sudden temperature changes that again punctuated the last glaciation. Several times, temperatures leaped up by 3.6 to 18°F within a decade or so, before recovering after a few hundred years. So far, more than twenty of these warm phases have been identified in the ice-core record. During many of them, temperatures in Europe at least may have been as warm as today.

These warming events, too, seemed to have some kind of periodicity or pulse. Temperatures moved from cold to warm and back again repeatedly, with a cycle ranging between 1,300 and 1,800 years. It was a recognizable pulse, just as a human pulse that races and then slows is recognizable, and averaged a full cycle roughly every 1,500 years. This pulse also swiftly got a name, the rather cumbersome Dansgaard-Oeschger cycle, after Dansgaard and his Swiss colleague, Hans Oeschger. Some interpret the data as showing a continuous background temperature cycle that on most but not all occasions triggered a more substantial warming episode during its warm phase, and on rather fewer occasions triggered a Heinrich event during its cold phase.

The connection between Heinrich events and the Dansgaard-Oeschger cycle wasn't recognized immediately—understandably enough. They had different time signatures, and one was revealed in the sediments of the mid-Atlantic, while the other emerged from the Greenland ice cores. Both, in any case, seemed at first to be minor local curiosities confined to the last glaciation, and therefore of no relevance to modern climate. But Bond had a hunch that the two were linked in some way, and that they had a global significance. Both, he noted, appeared to coincide with other climate changes, such as the advances and retreat of glaciers in Europe and North America. Like the Younger Dryas event and the climate flip 8,200 years ago, they seemed either to push the world into a different climate mode or to be part of such a process. Down the corridor, Bond's buddy Broecker was on hand to suggest a possible link to the ocean conveyor. The story began to take on a life of its own. But first the pair needed evidence to back up their hunch.

Bond began to re-examine trays of sediment cores from the bed of the North Atlantic that were assembled in his New York archive. Some were old cores, taken years before by the Lamont-Doherty research vessel *Vema* from beneath the waters off Ireland and the channel between Greenland and Iceland. Others were new, drilled off Newfoundland under Bond's supervision.

As expected, Bond found further evidence of Heinrich's rock fragments roughly every 8,000 years or so through the last glaciation. But the marine sediment cores also revealed lesser layers of materials normally alien to the seabed of the North Atlantic. Most exciting of all, these lesser layers occurred roughly every 1,500 years, and appeared to coincide with the cold phase of the Dansgaard-Oeschger cycle in the Greenland ice cores. This was pay dirt. Doubly so when it became clear that the iceberg armadas of the Heinrich events occurred during unusually cold phases of the Dansgaard-Oeschger cycle. The pattern seemed to involve a large Heinrich event, followed by five less and less severe 1,500-year Dansgaard-Oeschger cycles, and then another big Heinrich event. Sometimes this stately progression is influenced by other cycles, such as a solar precession, but otherwise it seems to hold.

Most remarkable of all, perhaps, Bond found that although there have been no Heinrich events during the 10,000 years since the end of the last ice age—the last was 15,000 years ago—the marine imprint of the underlying 1,500-year pulse has not missed a beat. "The oscillations carry on no matter what the state of the climate," he said.

Bond died in 2005, at the age of sixty-five. His longtime colleague Peter deMenocal has continued his work, looking for more signs of the pulse. Examining seabed sediments off Africa's west coast, he has found that every 1,500 years or so there were huge increases in dust particles in the sediments, suggesting big dust storms on land. The sediments also revealed dramatic increases in the remains of temperature-sensitive marine plankton, suggesting a temperature switchback in tropical Africa of as much as 9°F. "The transitions were sharp," deMenocal says. "Climate changes that we thought should take thousands of years to happen occurred within a generation or two."

Bond's final claim, that the pulse can be seen in recurrent climatic events right through to the present, seems to be vindicated, especially by temperatures in Europe and North America. There was an especially strong cooling event in the Northern Hemisphere that ended around 2,000 years ago; it was replaced by the medieval warm period that reached its height perhaps 1,100 years ago, and then by another cold era that bottomed out around 350 years ago, during the Maunder Minimum—when temperatures fell by up to 3.6°F in northern Europe, and the Eskimos reached Scotland in their kayaks.

Bond's study was an extraordinary piece of detective work. But it raises more questions than it answers. Two stand out. What, if any, is the relationship between these cycles and other parts of the climate system, such as Broecker's ocean conveyor? And, of course, what causes the mysterious pulse?

Heinrich originally argued that his ice armadas must be the result of some instability in the North American ice sheet that caused periodic collapses into the North Atlantic. There might thus be some link to big freshwater breakouts like that which triggered the Younger Dryas event. Certainly they involved huge amounts of ice. But the timing is fuzzy. Bond argued that while instabilities in the ice sheet could explain Heinrich events, only some of his pulses produced Heinrich events. So instability in ice sheets is unlikely to explain the pulses themselves, which in any case seem to have been unaffected by glaciations. By 2001, Bond believed he had confirmed the answer that many suspected all along.

He went back to the Greenland ice cores to look for evidence of solar cycles. There is no known direct marker for solar cycles in the cores. But other researchers had discovered that isotopic traces of cosmic rays bombarding the atmosphere were left in the ice cores—and that when solar radiation is at its most intense, cosmic rays are literally blown away from the solar system. Thus fewer "cosmogenic" isotopes, like carbon-14 and beryllium-10, are left in the ice cores during periods of strong solar radiation.

Bond came up trumps again. The evidence tallied. Over the past 12,000 years, fluctuations in detritus from the iceberg armadas in the Atlantic coincided with changes in cosmogenic isotopes in the ice cores. Thus there was a solar pulse that translated into a pulse in icebergs, global temperatures, and recurrent climatic events found through both the glacial and the postglacial eras.

Bond was convinced before his death that most climate change over the past 10,000 years had been driven by his solar pulse, amplified through feedbacks such as ice formation and the changing intensity of the ocean conveyor. He worried that people might interpret this as showing that global warming was natural. "But that would be a misuse of the data," he told me in an interview shortly before his death. Rather, he said, the most important lesson from his research is what it shows about the sensitivity of the system itself: "Earth's climate system is highly sensitive to extremely weak perturbations in the sun's energy output." And if it is sensitive to weak changes in solar forcing, it is likely to be sensitive also "to other forcings, such as those caused by human additions of greenhouse gases to the atmosphere."

What, exactly, drives the amplifications is another matter, however. For years, as Bond worked on his ideas, Broecker had declared that the Dansgaard-Oeschger temperature cycle in Greenland was linked to fluctuations in his ocean conveyor. Certainly the geography seemed right. Both appeared to originate in the far North Atlantic. It seemed clear, too, that during the periods when ice armadas were floating south in the Atlantic, temperatures in the North Atlantic were cold, and the amount of deep water being formed around Greenland declined. In extreme cases— perhaps during full-scale Heinrich events—the conveyor probably shut down. Perhaps a reduction in solar radiation triggered the entire sequence. But the evidence of what caused what was largely circumstantial. And as we will see later, there is another explanation, producing a large amplification from another quarter entirely.

But whatever the amplifier, the pulse is real and extremely pervasive. In the postglacial era, perhaps only in the past fifty years has something come along with greater power to disrupt climate.

VI

Tropical heat

26

THE FALL

The end of Africa's golden age

If there was a golden age for humans on Earth—a Garden of Eden that flowed with milk and honey—then it was the high point of the Holocene, the era that followed the end of the last ice age. From around 8,000 to around 5,500 years ago, the world was as warm as it is today, but there appear to have been few strong hurricanes and few disruptive El Niños; and it was certainly a world in which the regions occupied today by great deserts in Asia, the Americas, and especially Africa were much wetter than they are now. Optimists suggest that such conditions might await us in a greenhouse world. As we shall see, there are celestial reasons why that might not happen. But the Holocene era, and its abrupt end, may still offer important lessons about our future climate in the twenty-first century.

No place on Earth exemplifies the fall from this climatically blessed state better than the Sahara. The world's largest desert was not always so arid. Where seas of sand now shimmer in the sun, there were once vast lakes, swamps, and rivers. Lake Chad, which today covers a paltry few hundred square miles, was then a vast inland sea, dubbed Lake Megachad by scientists. It was the size of France, Spain, Germany, and the UK put together. Today, the lake evaporates in the desert sun; but then, it overflowed its inland basin and, at different times, drained south via Nigeria into the Atlantic Ocean, or east down a vast wadi to the Nile.

The difference is that back then, the Sahara had assured rains. The whole of North Africa was watered by a monsoon system rather like the one that keeps much of Asia wet today. Rain-bearing winds penetrated deep into the interior. From Senegal to the Horn of Africa, and from the shores of the Mediterranean to the threshold of the central African rainforest, vast rivers flowed for thousands of miles. Along their banks were swamps, forests, and verdant bush.

Beneath the Algerian desert, archaeologists have found the sand-choked remains of wadis that once drained some 600 miles from the Ahaggar Mountains into the Mediterranean. And in southern Libya, a region so waterless that even camel trains avoid it, archaeologists are finding the bones of crocodiles and hippos, elephants and antelope. If there was a vestige of true desert at the heart of North Africa, it was very much smaller than the desert is today. And, of course, there were people—shepherds and fishers and hunters—and some of the earliest known fields of grains like sorghum and millet. Archaeologists digging in the sands of northern Chad, currently the dustiest place on Earth, have found human settlements around the shores of the ancient Lake Megachad. Paintings in caves deep in the desert depict the lives of the inhabitants of the verdant Sahara of the Holocene.

There are other remains from this time. Rocks beneath the Sahara contain the largest underground reservoir of freshwater in the world. They were filled mostly by leaking wadis in the early Holocene. Some desert settlements today tap these waters at oases. Colonel Gadhafi has constructed pumps and a huge pipeline network to take this water from beneath southern Libya to his coastal farmers. He calls the network his Great Man-made River, though it is a feeble imitation of the real rivers that once ran here.

The wet Sahara and the era known more generally as the African Humid Period began around 13,000 years ago, as the ice age abated; and, except for the Younger Dryas hiatus, it lasted right through to the end of the golden age. It coincided with a time when Earth's precession ensured that the sun was blazing down on the Sahara with full intensity in summer. The land cooked, and convective air currents were strong. As the warm air rose, wet air was drawn in from over the Atlantic to replace it. The process was the same one that creates today's monsoon-rain system in Asia. Meanwhile, the monsoon rains were recycled by the rich vegetation across North Africa. Rather as in the Amazon today, the rain nurtured lush vegetation that ensured that much of it evaporated back into the air. The continually moistened winds took rain to the heart of the Sahara.

But the African Humid Period came to an end very suddenly. In the space of perhaps a century, the rivers of the Sahara emptied, the swamps dried up, the bush died, and the monsoon rain clouds were replaced by clouds of wind-blown sand. The climate system had crossed a threshold that triggered massive change. What happened? The first answer is that the sun moved. Or, rather, the precession continued its stately progress and gradually took away the extremely favorable conditions for Saharan rains. And as summer solar heating lessened, the warm air rose a little less and the monsoon winds from the ocean penetrated a little less far inland some years. The process was gradual, and went on without any appreciable effect on rainfall in most of the Sahara for more than 3,000 years. The vegetation feedback ensured that, at least in most years, the rain kept falling. If Lake Megachad was retreating, we have no evidence of it.

But at some point, the feedback began to falter. Perhaps there was a chance variation in rainfall that dried out the bush for a year or two. The sun was no longer strong enough to make good and revive the rains. Suddenly, what had been a feedback that kept the Sahara watered became a feedback that dried it out. The system as a whole had passed a threshold, and it never recovered. The green Sahara had become a brown Sahara. The North African monsoon rains had died.

Not everybody agrees that the vegetation feedback was the only trigger for the drying of the Sahara. One of Gerard Bond's solar pulses may have had some influence. But climate models show that in all probability, this flip in the Saharan climate was extremely sudden. Martin Claussen, of the Potsdam Institute for Climate Impact Research, in Germany, has played out this tragedy in detail in his model. He turns time forward and backward, recreates the subtle orbital changes, and fine-tunes the vegetation feedbacks. More or less whatever he does to mimic the conditions of 5,500 years ago, the result is the same. The system flips abruptly, turning bush to desert, and seas of water to seas of sand.

Other researchers have replicated his findings. Peter deMenocal, of Lamont-Doherty, calculates that the system flipped when solar radiation in the Sahara crossed a threshold of 470 watts per 10.8 square feet. Jon Foley, of the University of Wisconsin, found that a reduction in Holocene summer sun sufficient to reduce temperatures by just 0.72°F would have cut rainfall across the Sahara by a quarter, and by much more in the farthest interior of the continent. He says that once a region like the Sahara becomes dry and brown, it requires exceptional rains to break the feedback and trigger a regreening. Beyond a certain point—such as that reached 5,500 years ago—virtually no amount of extra rain is likely to be enough. The lack of vegetation "acts to lock in and reinforce the drought."

Back then, the people of the Sahara couldn't have known whether the droughts that suddenly afflicted them were permanent or not. But as the desert asserted control across the region, and the lakes and waterways dried up, they had no alternative but to leave. As part of the exodus, lakeside settlements near the Sudanese border in Egypt were all abandoned at about the same time. One was Nabta, famous now as the site of the world's earliest known stone structures with an astronomical purpose. They predate Stonehenge, in England, by about a thousand years. The key stones point to where the sun would have set at the summer solstice 6,000 years ago. Beneath some of the stones are burial sites for the cattle that the people tended. Nobody can be sure what the precise purpose of the structures was, but it is intriguing to suppose that they were used in an attempt to track the celestial changes that were disrupting the rains and turning their pastures to desert.

It may have been from such places that the myths and legends of past golden ages, and of the Garden of Eden, first emerged. The people who departed from the Sahara to set up new homes on the Nile or even farther afield would have taken their memories of a golden past. Researchers who have tried to date events in the Bible calculate mankind's expulsion from the Garden of Eden at around 6,000 years ago, when kingdoms across the Sahara would have been collapsing. But the Garden of Eden need not have been in the Sahara, because similar stories were played out elsewhere. Arabia dried out at the same time, leaving behind a huge underground reservoir of water not much smaller than that beneath the Sahara. Claussen calculates that the desertification of Arabia could have been caused by the same combination of gradual orbital change and a dramatic vegetation feedback.

The evidence is as yet sketchy, but the dramatic drying of the Sahara and Arabia appears to coincide with other climate changes around the world. In the Pacific Ocean, El Niño appeared to switch into a more active mode at around this time. There were cold periods from the Andes to the European Alps. In both cases, glaciers advanced strongly down their valleys; many of them are only today returning to their former positions. In the Austrian Tyrol, one victim of the advance was the "ice man" named Otzi, whose freeze-dried remains emerged from melting ice in 1991. In Ireland, a 7,000-year temperature record held in tree rings shows a cold era that included the coldest summers in the entire record, at about this time.

All this is particularly intriguing because—unlike during previous great climatic events of the era of the ice ages—there is little evidence that the primary action had much to do with the polar regions. It seems to have been an abrupt climate change formed in the tropics, with its major impacts there, and only ripples beyond. One in the eye for Wally Broecker, some of its investigators have been heard to say—a point to which we will return.

But what does this say about the future of the Sahara? Could warming in the twenty-first century trigger a greener, wetter Sahara? It is an intriguing idea, with plenty of adherents. Reindert Haarsma, a climate modeler at the Royal Netherlands Meteorological Institute, says the Sahara could be destined for a 50 percent increase in rainfall—enough to trigger a return to the golden age, in which crocodiles floated through swamps where today locusts swarm. Claussen, whose model first stimulated the idea, is more skeptical. He points out that the orbital situation now is very different, so summer solar radiation is not great enough to create a revived African monsoon. DeMenocal says solar radiation is currently 4 percent lower in the Sahara than it was when the Holocene flip occurred. But on the other hand, he admits, much higher levels of carbon dioxide in the air might compensate for this by stimulating an earlier recovery of Sahara vegetation.

Optimists point out that on a very modest scale, something of a revival is going on in Saharan rains and vegetation—albeit from the depths of the droughts that afflicted the region in the 1970s and 1980s. It hasn't happened everywhere, and some places have since slipped back. But, according to Chris Reij, of the Free University, in Amsterdam, improved farming methods, such as digging terraces and holding water on the land, may have

encouraged a modest greening of parts of the Sahara, and the resulting vegetation feedback could be one reason for the revived rains. But it would be
a big step to predict from that a reversion to the "Garden of Eden" days.

While some in the Sahara may conceivably be able to look forward to
greener, wetter times, the prognosis for many other arid regions around
the world is not so good. The big fear, from the American West to northern China, and from southern Africa to the Mediterranean, is of a twenty-
first century dominated by longer and fiercer droughts.

Again, history is the first guide. DeMenocal has been looking at the
history of droughts and civilization in the Americas, and finds strong evidence of periods of drought much longer than any known in modern
times. "There is good scientific evidence that vast regions of North America witnessed several such periods during the last millennium, with devastating cultural consequences," he says. "These megadroughts can persist
for a century or more."

The six-year Dust Bowl of the 1930s, which caused mass migrations
westward, was "pale by comparison" with its predecessors. Droughts in the
nineteenth century devastated many Native Americans as well as their bison. At the end of the sixteenth century, a twenty-two-year drought destroyed an early English colony at Roanoke, in Virginia. It became known
as the Lost Colony after all its inhabitants disappeared between their ar-
rival, in 1587, and the return of a supply ship four years later. Going back
earlier, tree rings show there was near permanent drought from 900 to
1300 west of the Mississippi and through Central America, which destroyed the Mayan and Anasazi civilizations. DeMenocal concludes that
complex, organized societies can get by in short droughts. They have
stocks of food and water, and know how to trade their way out of trouble
in the short term. But few of them can deal with megadroughts. If hunger
doesn't get them, the strife and turmoil caused by trying to survive does.

And the signs are that worsening droughts are becoming the norm in
regions that have suffered megadroughts in the past. In the American
West, the biggest river, the Colorado, is a shadow of its former self. Early
in the twentieth century, the average flow was 13 million acre-feet a year.
From 1999 to 2003, the average sank to 7 million acre-feet—worse even

than the Dust Bowl years. In 2002, it fell to just 3 million acre-feet. In 2005, the drought was continuing. In Central Asia, the Afghan war of 2002 was fought against a backdrop of drought as debilitating as any Taliban tyranny. The Hamoun wetland, which covers 1,500 square miles on the remote border between Afghanistan and Iran, has for millennia been a place of refuge for people from both countries in times of trouble. But that year it dried out and turned to salt flats. The water has not returned. Southern Europe is increasingly beset by forest fires and desiccated crops.

Richard Seager, of Lamont-Doherty, says that there is a long-standing correlation between drought in the western U.S. and drought in South America, parts of Europe, and Central Asia. And that is a pattern we see reasserting itself in the twenty-first century, as the Arizona desert creeps north, southern Europe increasingly resembles North Africa, and Central Asia takes on the appearance of Iraq or the Arabian Peninsula. Kevin Trenberth, of the National Center for Atmospheric Research, reports that the percentage of Earth's land area stricken by serious drought has more than doubled in thirty years. In the 1970s, less than 15 percent of the land was drought-stricken, but by the first years of the twenty-first century, around 30 percent was. "The climate models predict increased drying over most land areas," he says. "Our analyses suggest that this may already have begun."

That seems to be a common view. Mark Cane, a specialist in Pacific weather at Lamont-Doherty, says scarily: "The medieval warm period a thousand years ago was a very small forcing compared to what is going on with global warming now. But it was still strong enough to cause a 300- to 400-year drought in the western U.S. That could be an analogue for what will happen under anthropogenic warming. If the mechanisms we think work hold true, then we'll get big droughts in the West again." The Garden of Eden it is not.

Many believe that El Niño and the pattern of ocean temperatures in the Pacific are heavily implicated in the historical megadroughts, perhaps as part of a global reorganization of climate systems linked to Gerard Bond's pulses. And this should set modern alarm bells ringing, says Ed Cook, a leading tree-ring expert at Lamont-Doherty: "If warming over the tropical Pacific promotes drought over the western U.S. . . . any trend toward warmer temperatures could lead to a serious long-term increase in aridity over western North America." Martin Hoerling, of the National Oceanic and Atmospheric Administration, thinks that such a process is already under way. He blames the increasing droughtiness of the tropics on a persistent ocean warming in the Pacific that, he says, is "unsurpassed during the twentieth century." The pattern of dryness is beginning to look less like a local, short-term aberration and more like a long-term trend, he says, and he predicts that global warming "may be a harbinger of future severe and extensive droughts."

It won't happen everywhere, of course. Climate models predict that a warmer world will, on average, have more moisture in the atmosphere, and that, in general, the wet places will get wetter and the dry places will get drier. They predict that areas of uplift, where rising air will trigger storm clouds and abundant rain, will see the uplift become more intense. But areas of sinking air, which are the traditional desert lands of the world, will see more-intense sinking and drying. In many parts of the world, this "hyperweather" is likely to set competing forces against each other. Stronger storms will blow off the oceans, and monsoon-type rains may begin again in some places. But the rain-bearing winds will often be confronted by intensifying arid zones of descending air in the continental interiors. It is not obvious which force will win, and where.

Will the Sahara Desert expand and intensify, as drought theorists argue? Or will North Africa be reclaimed by a revived African monsoon? Megadrought or Garden of Eden? Nobody can answer that question yet. Perhaps the greatest likelihood is that in many places, from the Sahara to the American West and Arabia, there will be more and longer droughts, interspersed with brief but devastating outbreaks of intense storms and floods.

27

SEESAW ACROSS THE OCEAN

How the Sahara Desert greens the Amazon

Two of the world's largest and most fragile ecosystems face each other across the Atlantic. On one side is the Amazon rainforest; on the other the Sahara. They seem to be ecological opposites, and unconnected. The Sahara is rainless and largely empty of vegetation. The Amazon is one of the wettest places on Earth, and certainly the most biologically diverse, with perhaps half of the world's species beneath its canopy. But these two opposites are not so far apart. For one thing, the physical gap is surprisingly small. The Atlantic is narrow near the equator, and the two ecosystems are less than half as far apart as London and New York. For another, many believe they have a surprising symbiosis. Their fates may be intertwined in a rather unexpected way—and one that could have important consequences in the coming decades.

The key to the symbiosis lies in the remote heart of the Sahara, a region called Bodele, in northern Chad. Few people go here. It is littered with unexploded bombs and land mines left behind during Libya's invasion of Chad in the 1980s. And it is by far the dustiest place on Earth. Satellite images show year-round dust storms raging across Bodele and entering the atmospheric circulation. According to Richard Washington, of Oxford University, two fifths of the dust in the global atmosphere comes from the Sahara, and half of that comes from Bodele.

Some of this dust stays local. But much of it is carried on the prevailing winds, which cross the desert wastes of Niger, Mali, and Mauritania before heading out across the Atlantic. The red dust clouds can grow almost 2 miles high as they approach America. They cause spectacular sunrises over Miami, before falling in the rains of the Caribbean and the

Amazon. And there have been a lot of good sunrises in recent decades. The amount of dust crossing the Atlantic grew fivefold between the wet 1960s and the dry 1980s.

The Sahara dust has a series of unexpected effects on the Americas. According to hurricane forecasters in Florida, during dry, dusty years in the Sahara, there are fewer hurricanes on the other side of the Atlantic. It seems that dust in the air interrupts the critical updrafts of warm, moist air that fuel the storms. Equally surprisingly, desert bacteria caught up in the winds are being blamed for bringing new diseases to Caribbean coral reefs, and even for triggering asthma among Caribbean children.

And there is another important link. Saharan dust storms carry huge amounts of minerals and organic matter that enrich soils widely in the Americas. Bodele dust seems especially valuable. Its dunes are the dried-out remains of the bed of the vast Lake Megachad, which covered the central Sahara until its abrupt demise. Most of the dunes are made not of sand or broken rock but of the remains of trillions of diatoms, microscopic freshwater creatures that once lived in huge numbers in the lake. These fragments blow freely in the wind. That's why they make such plentiful dust storms. And they also make great fertilizer. If Bodele had any rain, the diatoms would make rich farmland. Instead, Chad's loss is the Americas' gain, says Hans Joachim Schellnhuber, a German physicist turned Earth-system scientist, who, as director of Britain's Tyndall Climate Centre, in Norwich, has made a study of the unlikely connection. "Bizarre as it may seem, the arid, barren Sahara fertilizes the Amazon rainforest. This process has been going on for thousands of years, and is one reason why the Amazon basin teems with life."

The two unique habitats are on a kind of seesaw, he says. When the Sahara is dry, as it has been for much of the past quarter century, its dust crosses the Atlantic in huge quantities and fertilizes the Amazon, making the rainforest superabundant. When the Sahara is wet, the dust storms subside and the Amazon goes hungry. That the Sahara seems to have only two basic modes, wet and dry, suggests that there may be two distinct modes in the Amazon, too. The last big change in the Sahara came 5,500 years ago, when the region lurched from wet to dry, probably within a few decades. As yet we know little about how the Amazon changed at that time. But if Schellnhuber is right, the Sahara's loss at that time may have been the Amazon's gain. There may have been a major change for the better in the rainforests.

In the twenty-first century, the seesaw could be on the move again. There are hints that the Sahara may become wetter, says Schellnhuber. And if the wetting turns to greening, and the vegetation feedback kicks in, the whole of North Africa could change dramatically. That would be good news for the Sahara, of course. But it might be bad news for the Amazon, which already seems to be close to its own tipping point, as the climate dries and rainforests give up their carbon. Could a wetter Sahara be the final nail in the Amazon's coffin? Schellnhuber believes it could.

28

TROPICAL HIGH

Why an ice man is rewriting climate history

There are two special things about Lonnie Thompson. First, doctors reckon that he has spent more time on mountains above 20,000 feet than any other lowlander on the planet. And second, in his freezer back home in Columbus, Ohio, he has probably the most detailed physical record anywhere of the climate of planet Earth over the past 20,000 years. Not bad for the sixty-year-old son of a hick from Gassaway, a tiny railroad town in West Virginia.

Make that three things. Because Thompson is, in a mild-mannered but determined way, a revolutionary in the world of glaciology. For four decades now, climate scientists have been drilling ice cores in the polar regions to find the secrets of climates past. They have found a lot, and they have developed some impressive theories about how the world's climate system is driven from these cold wastelands. But thirty years ago, Thompson, then still a graduate student in the geology of coal with a temporary post drilling ice cores in Antarctica, set out to prove them wrong about the origins of climate.

With his early mentor, the legendary British glaciologist John Mercer, Thompson ignored the poles and began drilling ice cores in glaciers high in the Andes, the Himalayas, and other mountain regions of the tropics. This was unheard of at the time. Finding funding was hard, because nobody had a budget for such work. But in the years since, he has uncovered a new, entirely unexpected world of tropical climate change. And now, after fifty expeditions to five continents, and with 20,000 feet of ice cores stored in his freezer, he believes he is on the path to proving that the true triggers and drivers and Achilles heels and thresholds and tipping points for the world's climate lie in the tropics.

For men like Broecker, this is sacrilege. But although Thompson's case is not yet proven, he has found some unexpected fans. Richard Alley, a career member of the "polar school," is an admirer of the senior from Gassaway. He told me with a smile: "Lonnie is a legend, and he may well turn out to be right." Whether he is right or not, Thompson's ice cores and the data he has painstakingly extracted from them are the lifeblood of an emerging debate between the polar and tropical schools—a debate that might not be happening at all without him.

Thompson is a loner. He has always avoided the big organizations and funding bodies that dominate so much climate science. Sometimes that has been out of necessity; now he sees it as a virtue. It has given him the freedom to do and think things his way. With his researcher wife, Ellen Mosley-Thompson, he set up a small team at the Byrd Polar Research Institute, part of Ohio State University. "We started small and we try to be self-contained," he says. "That makes us flexible. We don't have to stand in line for analysis of cores, or for supplies. And we have our own workshops to make everything."

The Thompsons build their own lightweight drills and photovoltaic generators, because these are the only means of getting the right gear by horseback onto the high slopes of the world's tallest mountains. And they have their own four automatic mass spectrometers, working 24 hours a day 365 days a year to analyze the samples brought back from around the world. Thompson doesn't even trust the big science institutions to look after his ice cores when he's gone. With the prize money that has come his way in recent years, he has created a trust fund to keep the freezers going in perpetuity.

Being independent means he can pack his bags and head around the world on a whim if he thinks there is an ice core to be had. Back in 1997, he took advantage of a brief thaw in diplomatic relations between Moscow and Washington to fly to Franz Josef Land, in the Siberian Arctic. There he extracted a thousand feet of ice from near an old Russian nuclear bomber base, and persuaded the bomber pilots to fly it back to Moscow for him. More recently, after years of stonewalling by the Tanzanian authorities, he took his drilling kit on a tourist flight to Dar es Salaam and smooth-talked his way up Kilimanjaro to extract vital evidence of the demise of its ice cap.

Thompson has spent half a lifetime taking his ice pick, crampons, and drilling gear to the Andes and the Himalayas, Tibet and the Russian Arctic, Alaska and East Africa. Back in Columbus, he has interrogated the ice and the bubbles of air trapped inside for signs of dust, metals, salts, and isotopes of oxygen and carbon to discover not just temperatures and rainfall but the comings and goings of El Niño events, forest fires, droughts, and monsoons.

His first love, he says, is Quelccaya, the first ice cap he scaled in Peru with John Mercer. It is the one he keeps going back to. He can see the whole world evolve there, he says, from the revival of El Niños in the Pacific around 5,500 years ago to the decades of drought that finished off the pre-Columbian Moche empire; from the first record in the tropics of the little ice age to the recent isotopic signature of global warming. Here and elsewhere across the tropics, he has also found a dust "spike" in the ice that shows that dust storms were sweeping across the tropics 4,200 years ago—evidence, it seems, of a sudden near-global megadrought.

Most intriguing for glaciologists, Thompson's collection of worldwide ice cores has revealed a previously unknown pattern in the formation of glaciers across the tropics. The pattern seems to be independent of the great glaciations that waxed and waned in the polar regions. It seems instead to follow latitude, starting in the Southern Hemisphere close to the Tropic of Capricorn, where he has found evidence that glaciers began to form in Bolivia 25,000 years ago. Then, as if by clockwork, other glaciers began to form and grow farther north. One by one, they started through Peru and Ecuador. Then, 12,000 years ago, a continent to the east but following the same northward trajectory, an ice cap began to form at the summit of Kilimanjaro, on the equator. Skipping north again to the Himalayas, around 8,000 years ago, glaciers started to grow near the Tropic of Cancer. Across three continents, glacier formation was oblivious of longitude or the equator or anything else. Latitude ruled.

Why? Thompson has tied this extraordinary progression to the precession, the wobble in Earth's orbit that gradually alters the line of latitude where the most intense solar heating occurs. This is the same wobble that sustained the African monsoon over the Sahara when the sun was overhead there in the early Holocene, but snuffed out the rains as the sun moved on. In the mountains of the tropics, glaciers generally started where the sun was fiercest. The sun was most intense over the Tropic of Capricorn 25,000 years ago and then moved north, becoming most intense over the Tropic of Cancer. It appears to have triggered the formation of glaciers all the way.

On the face of it, this seems odd. Why would the harshest sun and hottest temperatures create glaciers? Thompson has a simple explanation. The zone of maximum sun in the tropics is also the zone of maximum rainfall, which in the highest mountains means the zone of maximum snow. Up there, he says, it has always been cold enough for glaciers to form. So temperature is not an issue. What the high valleys have often lacked is moisture to feed the growth of glaciers. The sun brought the moisture, and with it the snow and the glaciers.

Many would argue that all the natural variability in climate that Thompson is uncovering offers a soothing reminder that the planet and human society are no strangers to climate change. Not Lonnie. His analysis is uncovering invisible thresholds in the climate system, he says. Cross them, and the whole system goes into a spin, with dramatic cooling or warming, great droughts and the El Niño flip, turned full on or full off for centuries at a time. Should we not be just as concerned that carbon dioxide might send us above a threshold? If that happens, he says, "we won't get gradual climate change, as projected; we will instead get abrupt change."

And, of course, Thompson is tracking with concern the role of modern climate change in melting his glaciers. Back in 1976, he took a core of the ice at the summit of Quelccaya. It showed layers of ice laid down annually for 1,500 years. In 1991, when he returned to update the record, he found that the annual accumulation had stopped and the top 20 yards of ice had melted away—dramatic evidence of a recent and sudden shift in an ancient ice cap's fortunes. In the valley below, Quelccaya's largest glacier, the Qori Kalis, is retreating by 500 feet a year and has lost a fifth of its area since 1963. Across Peru, a quarter of the ice surface has disappeared in thirty years. Elsewhere in the Andes, Bolivia's Chacaltaya lost two thirds of its ice in the 1990s, and Venezuela has lost four of its six glaciers since 1975.

In Africa, where 80 percent of the ice on Mount Kilimanjaro has melted away in ninety years, Mount Kenya has lost seven of its eighteen glaciers since 1900; and most of the ice on the Rwenzori Mountains between Uganda and Congo has gone, too. Across the Indian Ocean, on New Guinea, the West Meren glacier vanished altogether in the late 1990s, and its neighbor Carstensz has shrunk by 80 percent in sixty years. Thompson has seen the same trends in the Himalayas and Tibet. Glacial retreat, he says, "is happening at virtually all the tropical glaciers." In some places, there may be local factors. Occasionally, declining snowfall will do the damage. But he insists that while snowfall in high altitudes may be critical to getting a glacier started, it is rarely critical to the glacier's demise, which starts lower down the slopes. Globally, he says, there can be no explanation for the universal disappearance of glaciers other than global warming.

Thompson believes that he has only begun to explore the potential of his ice cores to answer questions about the tropics. He wants to take cores from ice still attached to the Nevado del Ruiz volcano, in northern Colombia. The mountain exploded in 1985, engulfing 20,000 people in a landslide of ash. "I think we could get a record of how often that volcano erupts," he says, apparently oblivious of the risk for researchers in such an expedition. He believes that the ice of Quelccaya can offer a history of fires and drought in the nearby Amazon. And he is looking at dust from China that has collected in ice in Alaska. It is already providing a history of pesticide use in China, and may eventually reveal whether dust out of Asia, as well as that from the Sahara, could have fertilized the soils of the Americas.

Thompson believes that by uncovering the secret climate history of the tropics, he is helping to strip climatology of an unhealthy fixation with what happens close to the homes of the researchers—in the North Atlantic: "An important reason why we think that Greenland and those places are so important is because so much research has been done there— and that is mainly because it is more convenient than going to Tibet or Patagonia." He believes that that fixation is diverting researchers from where the real climatic action is—in the tropics, in the world of El Niño and the Asian monsoon and megadroughts and the dramatic feedbacks that dried up the Sahara, which he sees as "at least as important as anything Wally Broecker has cooked up on the North Atlantic."

To Thompson, it has always seemed obvious that "the global climate is driven from the tropics." Most of the surface of Earth is in the tropics, he says. "It is where the majority of the heat reaches Earth, and from where it is distributed around the globe. It is where the great climate systems like the monsoon and El Niño are based." He argues that truly global climatic events can start only where heat and moisture can be delivered both north and south around the globe. There may be feedbacks operating in the North Atlantic or around Antarctica. But the big drivers must be in the tropics.

Thompson has his own heroes. Mercer is one. Another is James Croll, the lowly Victorian Scot who worked his way through life as a waiter, a school caretaker, and a carpenter so that he could research the astronomical forces behind the ice ages. And Thompson has simple advice for young scientists: plow your own furrow. "Go somewhere or do something that nobody else has even thought about working on." Some academics from the wrong side of the tracks would have settled quietly into faculty life, thankful for their social advance. Not Lonnie. He does research the hard way. "On one trip we were up on Quelccaya for three months. We had to cut the ice cores by hand into 6,000 samples, take them downhill on our backs, and then melt them and put the water in bottles sealed with wax." On another occasion, he found himself in New Zealand dangling on a rope above 2,000 feet of empty space.

Years ago, a student in the field with Thompson died of the aftereffects of altitude sickness. His father sued. That still hurts. Thompson would be the last professor on Earth to send his students somewhere he wasn't prepared to go himself. He is still prepared to live for months under canvas in freezing cold and lung-achingly thin air. Just turning sixty when we met, he was recently back from his biannual trip to the Andes, and his calendar included upcoming trips to Kilimanjaro and central Africa's "mountains of the moon." He had tentative plans for expeditions to the last glaciers in

New Guinea and a Siberian island near where the last mastodon froze to death 5,000 years ago.

He told me he reckoned that his techniques could one day help uncover the remains of life in the ice caps of Mars. And I swear that his eyes lit up when I suggested he might be on the first flight to the red planet.

29

THE CURSE OF AKKAD

The strange revival of environmental determinism

Around 4,200 years ago, the world's top empire was run by Sargon, the despotic but otherwise unexceptional ruler of the Akkadian empire. Some have called this the first true empire in the world. Certainly it seemed to be a new form of society, created out of a number of previously autonomous city-states on the floodplains of the Tigris and Euphrates Rivers in Meso-potamia. Its rule extended all the way from the headwaters of the two rivers, in Turkey, across much of Syria and as far south as the Persian Gulf. But Sargon's empire had been in business for only a century or so when it suddenly collapsed. Archaeologists initially put this down to an invasion of barbarian hordes from the surrounding mountains. But an energetic field archaeologist called Harvey Weiss, of Yale, changed that rather lazy assumption—and with it changed much else about our perceptions of the rise and fall of past civilizations.

In the late 1970s, while working in Syria, Weiss discovered a "lost city" beneath the desert sands, close to the Iraqi border. Over more than a decade he excavated the remains of the settlement, named Tell Leilan. He pieced together the story of a highly organized city that had grown over several thousand years from a small village to a prosperous outpost of the Akkadian empire. But there was a mystery. It appeared that for some 300 years, the city had been abandoned and its streets had filled with wind-blown dust.

Weiss tied the events at Tell Leilan to a contemporary cuneiform text titled "The Curse of Akkad," which recorded a great drought in which the fields of most of northern Mesopotamia were abruptly abandoned. The granaries emptied, the fruit trees died in the orchards, and even the fish departed as the great rivers dried up. Refugees flooded south. The people of southern Mesopotamia built a hundred-mile wall to keep them out. Archaeologists had previously dismissed "The Curse of Akkad" as mythology. The idea that climatic and other environmental change determined the progress of societies had been hugely out of fashion. The prevailing view was that politics, economics, wars, and dynasties made and broke empires, and that climate was just a more or less benign backdrop.

But Weiss was convinced that only a massive shift in climate could explain a 300-year collapse, after which the climate apparently recovered enough for the northern plains to be settled once more. When he published his findings, they provoked consternation in the archaeological community but huge interest among climate scientists—not least Peter deMenocal, of Lamont-Doherty. "After Weiss's publication, environmental determinism had a huge revival," deMenocal says. Especially after it emerged that the dust storms of Mesopotamia were part of a wider process of aridification right across the Middle East and beyond, which had seen off other societies, too.

In New York, deMenocal was working with a student, Heidi Cullen, on analyzing a core of marine sediment drilled from beneath the Gulf of Oman, 1,500 miles south of Tell Leilan. They decided to look for evidence of dust storms in the core. "We thought the dust might be visible there, and Heidi started to go through it," he told me. "It was very painstaking work, and to be honest, she was about to give up. Then boom. One day she found it. The 300-year layer of dust, dated at 4,200 years ago, and much of it clearly derived from Mesopotamia. We sent it to Harvey, and he was ecstatic."

The news spread. Lonnie Thompson and his team went back to their tropical ice cores and found similar layers of black dust. "It was a huge global dust spike," he said. In the ice on the summit of Kilimanjaro, in East Africa, there is only one dust "spike" in the 12,000-year record. And it occurs right at 4,200 years ago, he said. On the other side of the planet from Syria, at Quelccaya, in Peru, the same period produced "the biggest dust event in the ice core in a 17,000-year record." Fallout of dust onto the glacier was a hundred times as much as normal levels. "And it shows up in the Asian monsoon region of the Himalayas, too," says Thompson's dust analyst, Mary Davis.

From Lake Van, in eastern Turkey, to the Dead Sea, in Palestine, and in Africa from Kenya to Morocco, water levels fell by tens or even hundreds of yards 4,200 years ago. Civilizations were ending everywhere. In Egypt, those years produced a collapse of order that marked the break between the Old and Middle Kingdoms. "On the tombs of the Pharaohs, their histories talk of expansion until 4,200 years before the present, when there were droughts and mass migrations and sand dunes crossing the Nile," says Thompson. In Palestine, the situation was even worse, according to Arie Issar, an Israeli hydrologist and the author of a detailed study of climate change and civilization in the region. The level of the Dead Sea dropped a hundred yards. "All the urban centers were abandoned, and the cities, which had existed for several hundred years, remained only as large heaps of ruins. They were not resettled until nearly half a millennium later." Farther east, in the Indus Valley of modern-day Pakistan, the urban centers of the Harappan civilization collapsed at the same time.

What caused all this? Nobody is sure. Jeffrey Severinghaus, of the Scripps Institution of Oceanography, has found tantalizing evidence of a dust signal in the Greenland ice cores 4,200 years ago. But instead of more dust than before, he found less. There was also a decline in sea ice in the North Atlantic. This has been interpreted as evidence of a change in the ocean conveyor. Did Broecker's conveyor drive things once again? On the face of it, that interpretation looks unlikely. For on this occasion, rather as during the great climatic disruption of 5,500 years ago, events in the North look like mere ripples flowing out from much bigger events in the tropics.

It is more evidence, says deMenocal, that climate switches may lurk in the tropics at least as much as at the poles. Richard Alley again reaches for common ground. Perhaps, he says, the Arctic feedbacks were at their height during and immediately after the ice ages, but lost their influence once most of the ice had gone. During the height of the Holocene, at least, perhaps the tropics ruled. But if so, what is driving the feedbacks in the tropics? Where are the tropical equivalents of Broecker's conveyor, Alley's "sink or freeze" switch, and Juergen Mienert's clathrate gun?

3 0

A CHUNK OF CORAL

Probing the hidden life of El Niño

Some researchers have a way of combining business with pleasure. Not for Dan Schrag, the Harvard geochemist, the arduous journeys into thin, cold air on tropical glaciers. Back in 1997, he was on his fourth trip to the paradise islands of the East Indies in search of ancient coral. One day, he was sauntering along a beach on Bunaken Island, a speck of old atoll off the Indonesian island of Sulawesi. "We had had a glorious dive, during which we saw a huge school of barracuda," he remembers. "We stopped for lunch, and I took a walk down the beach, behind the mangrove swamp. It was the last day of the trip. We had failed to find anything useful, and I was preparing to go home. Then I saw this massive coral head on the beach, incredibly well preserved." He chiseled out a piece and headed for the plane.

Back in the lab at Harvard, Schrag discovered that this fossilized piece of coral was 125,000 years old and contained sixty-five years' worth of growth rings that gave a brief window on the climate of the western Pacific back before the last ice age. It was a "fantastic discovery," he says. "I guess I got really, really lucky." The coral he had found was the first piece ever located that was large enough and well enough preserved to give a good snapshot of ancient El Niños. What's more, says Schrag, it came from a region that is in the "bull's-eye" of El Niño, in the heart of Indonesia. His preprandial discovery is helping transform our understanding of El Niño's place in the climate system.

Until recently, climatologists looked on El Niño as a minor aberration in the tropical Pacific, of only passing interest to the wider world. But in the past two decades it has become the fifth horseman of the Apocalypse, a bringer of devastating floods, fires, and famine from Ethiopia to Indonesia to Ecuador, and a sender of weird weather around the world. It has been appearing more frequently, and with effects that are much more violent and last longer. Its current level of activity is unparalleled in the historical record. Yet the historical record doesn't go back far, so nobody has been sure whether this is a perfectly normal upturn or an alarming consequence of global climate change. Schrag's coral has helped provide some answers. It makes a strong case that global warming is already having a profound effect on what climatologists are coming to regard as the flywheel of the world's climate.

El Niño is a periodic reversal of ocean currents, winds, and weather systems that stretches across the equatorial Pacific Ocean, halfway around the planet at its widest girth. It is a redistributor of heat and energy in the hottest part of the world's oceans, which kicks in when the regular circulation systems can no longer cope. In normal times, the winds and surface waters of the tropical Pacific, driven by Earth's rotation, flow from the Americas in the East to Indonesia in the West. In the tropical heat, the water warms as it goes. The result is the gradual accumulation of a pool of hot water on the ocean surface around Indonesia. This pool can be up to 13°F warmer than the water on the other side of the ocean, and can contain more heat energy than the entire atmosphere. All that heat generates storm clouds that keep the rainforests of Southeast Asia wet.

But the constant flow to the west also piles up water. Trapped against the Indonesian archipelago, the warm pool can rise as much as 15 inches above sea levels farther east. Clearly, this state of affairs cannot last. And every few years, usually when the winds slacken, this raised pool of warm water breaks out and flows back across the surface of the ocean, right along the equator. As the warm water moves east, the wind and weather systems that it creates follow.

Deprived of their storm-generating weather systems, Indonesia and a wide area of the western Pacific, including much of Australia, dry out. There are forest and bush fires, and crops shrivel in the fields. Meanwhile, the displaced wet and stormy rainforest climate drenches normally arid Pacific islands, and often reaches the coastal deserts of the Americas. Ripples from this vast movement of heat and moisture spread around the globe. They move west through the Indian Ocean, disrupting the Indian monsoon and causing rains or drought in Africa, depending on the season. They move east. Beyond the flooding on the Pacific shores of the Americas, El Niño brings drought in the Amazon rainforest. Its hidden hand alters flow down the River Nile, triggers rains in the hills of Palestine, and damps down hurricane formation in the North Atlantic.

Typically, an individual El Niño event lasts twelve to eighteen months. After it has abated, the system often goes into sharp reverse, with exceptionally wet conditions in Indonesia and fierce drought further east. This is called La Niña. Together, El Niño and its sister constitute a vast oscillation of ocean and atmosphere that in recent times has been the most intense fluctuation in the world's climate system.

Scientists first became aware of the oscillation we now call El Niño in the nineteenth century. But they have been uncertain about how far back El Niño goes. Is it a permanent feature of the climate system, or a minor and occasional aberration? Does it have long-term variability tied to global climate changes? Does the Pacific get "stuck" in either a permanent El Niño or a permanent La Niña?

Reliable climate and ocean records cover only a couple of centuries or so. Delving further requires alternative sources of information. To this end, Donald Rodbell, of Union College, in Schenectady, New York, dug up the bed of a lake in southern Ecuador to chart its past flood levels, in the expectation that, as today, floods would be a feature of El Niño episodes. In 1998, he published a remarkable 12,000-year record of the lake's floods. For the first half of the period, they came roughly once every fifteen years, suggesting a near-dormant El Niño. Then they speeded up quite abruptly, to settle at an average return period of about six years—the classic El Niño pattern until recently. This pattern has been confirmed by Lonnie Thompson's ice cores from nearby glaciers.

The change in the flood pattern also seems to coincide with the same precession shift in Earth's tilt that caused the desertification of the Sahara and the advance of tropical glaciers spotted by Thompson. Rodbell's record was a major breakthrough, implicating El Niño as a key driver of the global climate system. El Niño was no longer just a short-term cycle played out over a few months in one ocean: it had global and long-term meaning. Then came Schrag's chunk of coral.

Through isotopic analysis, Schrag extracted an El Niño signal from his piece of jetsam. When water evaporates, molecules containing the lighter isotope of oxygen—oxygen-16—evaporate slightly faster, leaving behind seawater that is rich in the heavier oxygen-18. When it rains, the oxygen-16 is returned. So in the Indonesian islands during El Niños, when rainfall ceases, both the seawater and the coral growth in those years contain more oxygen-18. Schrag measured the ratio of the two oxygen isotopes in the sixty-five annual growth rings in his ancient chunk of coral. He found two things of importance: First, there had indeed been an El Niño cycle back then. That pushed the longevity of the phenomenon back to before the last ice age, further establishing it as a permanent feature of the climate system. And second, the El Niño cycle looked exactly like that of the modern period from the mid-1800s to the mid-1970s, in which El Niño returned, on average, about every six years. This underlined the idea that six years is the natural length of the cycle—and made the post-1976 period, during which El Niño has developed a return period averaging 3.5 years, appear increasingly unusual.

This sense that El Niño may have changed in some fundamental way in the past thirty years has been reinforced by another change. The earliest records of the El Niño phenomenon are from the Pacific shores of South America, where a cold ocean current normally works its way north, bringing waters rich in nutrients that sustain one of the world's largest fisheries, off Peru. But during El Niños, the flood of warm water from the west overrides this cold current for a while, and the fish disappear. That has been the classic pattern. But since 1976, the underlying state has changed. The cold current has been pushed to ever-greater depths, even during normal times. The ocean system appears to have become stuck in a quasi-El Niño state.

What lies behind these recent changes? Some say that El Niño is simply on a short-lived, exuberant joyride. They point out that there have always been decades when it is unusually quiet or busy or just plain weird. But Schrag thinks this is unlikely to explain recent events. Publishing his Sulawesi findings, he said: "In 1982–83 we experience the most severe El Niño of the 20th century. According to previous records you wouldn't expect another that powerful for a hundred years. But 15 years later, in 1997–98, we have one even larger." And since then, in 2002 and 2004, there have been two more significant El Niños—not as large as those before, but turning up with ever-greater frequency.

Kevin Trenberth, the head of climate analysis at NCAR, was one of the first researchers to claim that the Pacific entered an unusual state after 1976. He believes that the recent spate of strong and frequent El Niños could well be due to the hand of man. It looks as if global warming, which gathered real pace only in the 1970s, is generating so much warming in the tropical Pacific that the old flywheel pattern in which occasional El Niños distribute the heat that accumulates around Indonesia is not sufficient to handle the amount of energy in the system.

Modelers have been testing this theory, with interesting results. Mojib Latif, at the Max Planck Institute for Meteorology, in Hamburg, developed the first global climate model that was detailed enough to reproduce El Niño. His model predicts that the average climate in the twenty-first century will be more like the typical El Niño conditions of the twentieth century. Cold La Niña events will still happen occasionally, and may even be more intense. But they will become the breakout events.

It would be wrong to suggest that science has somehow cracked the enigma of El Niño. There are still many mysteries. Certainly the idea that a strong El Niño is necessarily associated with warm times could be a gross simplification. Schrag's coral, along with other evidence, suggests that El Niño kept going right through the last ice age, when, even in the western Pacific, temperatures were several degrees lower than they are today. There is even some suggestion that El Niños were more common in the colder phases of the ice age, whereas La Niña held sway during the warmer periods. Likewise, the warm early Holocene era, before 6,000 years ago, saw El Niño largely in abeyance. It recovered during a cooler period.

Clearly, El Niño is not a simple planetary thermostat. But its operation in the past may have had more to do with changes in solar radiation that were reflected in alterations to the tropical hydrological cycle than with temperature. It is possible to imagine a climate system in which those changes triggered different temperature signals at different times. So efforts to tie past El Niños to temperature trends may not provide a good guide to what happens in a world of pumped-up greenhouse gas concentrations.

But what is becoming clear is that El Niño is a phenomenon that influences basic planetary processes such as the transfer of heat and moisture in huge swaths of the tropics. That it has big swings that operate on timescales varying from months to thousands of years. That it leaves its calling card in different ways right around the planet. And that its variability seems to be keyed into critical external drivers of past climate such as the precession and Bond's 1,500-year solar cycles. You would not bet against its playing an equally important role in moderating or amplifying global warming caused by greenhouse gases. What is not yet clear is which way it will jump.

Perhaps scientists should put aside their models and search for some wisdom on El Niño from Peruvian farmers, who have grown potatoes high in the Andes for thousands of years. Throughout that time, El Niños have become stronger and weaker, more frequent and less frequent, and have influenced potato growing all the while. For many centuries now, farmers have gathered in mid-June (the Southern winter) to gaze up at the night sky in the Andes and observe the eleven-star constellation known as Pleiades, or the Seven Sisters. If the stars are bright, they set to planting quickly, confident that there will be good rains and a healthy harvest.

For years this folklore was dismissed by agriculturalists as mumbo-jumbo—until Mark Cane, one of the world's foremost modelers of El Niño, heard about it from a guide while traveling in the Andes. Intrigued, he checked meteorological records, and discovered that typically about six months before an El Niño, thin, high, and almost invisible clouds form above the Andes. These dim the brightness of the constellation. So a dim constellation means a dry spring, while clear skies and a bright constellation mean good rains.

The farmers had thus perfected many hundreds of years ago what climate modelers like Cane have only fitfully managed in the past twenty years—a way of forecasting El Niño. Cane says the Peruvian potato farmers' forecast is better than his. "It's a brilliant scheme, really quite a feat. I still wonder how they possibly worked it out." Perhaps, he muses, the Peruvian potato farmers have had the key to the world's climate all along.

3 1

FEEDING ASIA

What happens if the monsoon falters?

More than 3 billion people today are fed and watered thanks to the Asian monsoon. It is the greatest rainmaking machine on the planet—and possibly one of the most sensitive to climate change. Its mechanism is extremely simple. It is like a giant sea breeze operating over the world's largest continent. In winter, the vast Asian landmass becomes cold—extremely cold on the high ice caps of Tibet, the largest area of ice outside the polar regions. It cools the air above. That air descends, forcing cold, dry winds to blow off the land and out across the Indian and Pacific Oceans. Asia is mostly rainless for nine long months. But come summer, the land warms up much faster than the oceans. Warm air rises, and as it does, the winds reverse and moist winds blow in off the oceans. For about a hundred days, monsoon rains fall across Asia. The rains burst rivers, fill irrigation canals and water fields. Across the continent, rice farmers take their opportunity to grow the food that sustains half the world's population.

A failed monsoon has devastating consequences. They happened repeatedly in the nineteenth century. British colonial administrators in India watched in bemusement as tens of millions died in the famine of 1837–1838, and again in 1860–1861, 1876–1878, and 1896–1902. The Asian monsoon remained an unruly beast through the twentieth century. But despite tenfold increases in the populations of most monsoon countries, the death toll from famine has fallen. There are many reasons for this, one of which is that the rains proved more reliable in the twentieth century than in the nineteenth. That was a good news story. The question is: Can it last? The Asian monsoon has appeared for the past century to be self-contained and invulnerable. But, like other big features in the global climate system, it may have an Achilles heel. If the monsoon proves less reliable in the twenty-first century, there could be real trouble ahead—for about 3 billion people.

The monsoon's vulnerability in past centuries seems to lie in its links to two other parts of the global climate system. One is El Niño. Strong El Niños often seem to switch off the Asian monsoon. British imperial scientists discovered more than a century ago that most of the great Indian famines coincided with marked climatic fluctuations in the Pacific. El Niños seemed to draw heat away from Asia, and so to drain the monsoon's strength. But the argument has become a little academic in the past thirty years, because El Niño has intensified without any widespread weakening of the Asian monsoon. The break in the old link has been both a scientific surprise and a humanitarian godsend. But nobody knows what has caused it and whether it will last. If the Pacific climate system does what many predict, and in the twenty-first century leans heavily toward a permanent El Niño–like state, and if the monsoon resumes its former relationship, then the rains may soon fail over Asia more often than they succeed.

The second link is with the Atlantic. This was dramatically established in 2003, when Indian and U.S. researchers assembled a 10,000-year record of the strength of the Indian monsoon. They did it by counting fossilized plankton found in ancient marine sediments off the Indian coast. The plankton thrive when strong monsoon winds cause an upwelling of the nutrients that provide their food. The study found huge variability in the monsoon's strength over the centuries. And it confirmed that, over time-scales longer than individual El Niño years, "weak summer monsoons coincide with cold spells thousands of miles away in the North Atlantic," according to Anil Gupta, of the Indian Institute of Technology, in Kharagpur, who worked on the project. Strong monsoons go hand in hand with warm waters off Europe and North America.

It had been known for a while that the Indian monsoon turned off during the last ice age but probably flickered on briefly during the warm episodes that punctuated the glaciation. The new study showed that the strength of the monsoon also shadowed the flutters of the Atlantic system during the postglacial era, faltering during the Younger Dryas and the chill of 8,200 years ago, for instance. The changes clearly followed Bond's

1,500-year solar pulse. Thus the last faltering of the monsoon came during Europe's little ice age, which ended in the final decades of the nineteenth century. Soon, as colonial records confirm, the monsoon was regaining its reliability.

But this pattern, impressive though it is, does not explain how the link with the Atlantic works. Does the Atlantic tell the monsoon what to do? Does the monsoon tell the Atlantic what to do? Does Bond's solar pulse independently determine both? Or is there another element not taken into account? Where does El Niño fit in, for instance?

Jonathan Overpeck, of the University of Arizona, one of the authors of the monsoon history, holds that the Atlantic has the whip hand. He says that a warm North Atlantic sends heat east on the winds, warming Asia in spring, and allowing a rapid melt of the Tibetan plateau and an early start to the rain-giving monsoon winds. But when the Atlantic is cold, he says, "more snow on the Tibetan plateau in spring and early summer uses up all the sun's heating, because it has to be melted and evaporated before the land can warm." If he is right, then should the ocean conveyor falter in the coming years, the effects for Asia could be even more grievous than for Europe. "There could be a weakened monsoon and less water for all the people who depend on it," says Overpeck.

The tropical school disagrees with this analysis. It holds that both the cooling of the Atlantic and the weakening of the monsoon are likely to be triggered by changes in the heating of the tropics. According to this theory, a cooling of the tropics will weaken monsoon winds and rains, while at the same time sending less warm water north in the Gulf Stream. The theories of the polar and tropical schools are on this occasion not mutually exclusive. In fact, they are mutually reinforcing.

But right now, neither theory offers much enlightenment about what might happen to the Asian monsoon in the coming decades. Global warming driven by accumulating greenhouse gases without a solar component may have different features and different outcomes from the solar-dominated scenarios of the past. The situation is further complicated because across much of monsoon Asia, warming is itself severely compromised and sometimes extinguished by the aerosols in the Asian brown haze. As we have seen, the haze's biggest impact is on the radiation balance between the land surface and the air aloft—a vital parameter in determining the strength of the monsoon. The fear is that the haze may break the seasonal heating cycles between land and ocean, and turn off the monsoon. It hasn't yet, but it may. And, valuable though reconstructed histories of the Asian monsoon may be, it is unlikely that they will ever be able to provide a firm prognosis for the monsoon.

VII

AT THE MILLENNIUM

32

THE HEAT WAVE

The year Europe felt the heat of global warming

At a zoo near Versailles, outside Paris, keepers kept twenty-seven polar bears cool by feeding them mackerel-flavored ice. In Alsace, the electricity company trained water cannons on the roof of a nuclear power reactor as temperatures outside soared to 118°F. In Rome, tourists queued up to pay the fine for bathing in Trevi Fountain. It seemed like a good deal, they said. Crops died; forests burned; power blacked out as office air conditioners were turned to full power; rivers from the Danube to the Po and the Rhine to the Rhone were at or near record lows.

This was by no standards an ordinary summer heat wave. For one thing, it killed at least 35,000 people: 20,000 in Italy and 15,000 in France. Old people, many of them abandoned in apartments without air conditioning as their families took their August holidays, suffered most. Dehydrated and short of breath, they died by the thousands in temperatures that often exceeded 104°F during the day and stuck close to 86°F at night. It was Europe's hottest summer in at least half a millennium. At the heat wave's peak, on August 13, 2003, the twenty-four-hour death toll in Paris was eight times the norm. In parts of the city, there was a three-week wait for funerals. More than 400 bodies were never claimed by relatives.

It wasn't just the mortuaries that were rewriting their record books. This was the first single weather event that climate scientists felt prepared to say was directly attributable to man-made climate change. In the past, the assumption had always been that any individual weather event could be the product of chance. But the 2003 heat wave was different, says the Oxford University climate scientist and statistician Myles Allen. "The immediate cause, I agree, was a series of anticyclones over Europe. They always raise temperatures in summer, and we can't say those were made any more likely by climate change. But we can say that climate change made the background temperatures within which those anticyclones operated that much higher."

There is no doubt that average temperatures have been rising strongly for years. In much of Europe, the summer average at the start of the new century was 0.9 to 1.8°F warmer than it was in the first half of the twentieth century. In the summer of 2003, temperatures averaged 4.1 degrees warmer. Judging from past averages, the heat wave was probably a once-in-a-thousand-years event. But, says Allen, "small changes in average temperatures make extreme events much more likely."

One of the nicest confirmations of how exceptional the summer of 2003 was came from a study published at the end of 2004. The French mathematician Pascal Yiou, of the Laboratoire des Sciences du Climat et de l'Environnement, had collected more than 600 years' worth of parish records showing when the Pinot Noir grape harvest began in the Burgundy vineyards of eastern France. There is a clear relationship between summer temperatures and the start of the harvest, so he extrapolated backward to produce a temperature graph from the present to 1370. The results showed that temperatures as high as those typical in the 1990s were unusual, but had happened several times before. "However," Yiou said, "the summer of 2003 appears to have been extraordinary, unique." Temperatures in Burgundy that year were almost 11°F above the long-term average. And if Yiou's formula was accurate, the highest previous temperature had been just 7° above the average. That happened in 1523, in a warm interlude during the little ice age.

"The 2003 heat wave was far outside the range of normal climate," says Allen. It was not impossible that it could have happened without global warming, but it was very improbable. "Our best estimates suggest the risk of such a heat wave has increased between four- and sixfold as a result of climate change." Many scientists continue to argue about how we might recognize "dangerous" climate change, he told me. "Well, for the thousands of victims in Europe in the summer of 2003, it is clear we have already passed that threshold."

And the big heat is only just beginning. Allen says that by midcentury, if current warming trends persist, the extreme temperatures experienced in 2003 in Europe could occur on average once every two years. Richard Betts, of Britain's Hadley Centre, says that for people living in cities, the risks are even greater. They are already feeling the worst of climate change, because they also suffer the "urban heat island effect." During heat waves, the concrete, bricks, and asphalt of buildings and roads hold on to heat much better than does the natural landscape in the countryside. In the typically windless, anticyclonic conditions of a European heat wave, the effect is even more marked. The air just stays in the streets and cooks. The effect is especially marked at night, which doctors say is a critical time for the human body to recover from daytime heat.

Betts says global warming will push the urban heat island effect into overdrive. Doubling carbon dioxide levels in the air will triple the effect, he calculates.

33

THE HOCKEY STICK

Why now really is different

It was a seductive image. So seductive that the IPCC put it right at the front of its thousand-page assessment of climate change, published in 2001. The panel hoped that it would become as talked about as the Keeling curve. And scientists gave it a snappy caption: this was the graph they called the "hockey stick." As I don't play hockey, I was initially left wondering why. But if you lay a hockey stick on the ground and look at its shape as a graph, you will see that the long, flat shaft has at the end of it a short but sharply upturned blade, the bit you hit the puck with. And that, according to the IPCC authors, is the shape of the world's temperatures over the past thousand years: about 900 years of little or no change, followed by a century with a short, sharp upturn.

The assembly of the data behind the hockey stick graph has become a political cause célèbre. It began with high hopes: it was to be the first serious attempt to piece together a global picture of climate over the past millennium from a wide variety of different kinds of sources. Rather than carrying on the well-established work of reconstructing past temperatures from analysis of tree rings, it sought to add in other proxy data from ice cores, coral growth rings, and lake sediments. The idea was to lose the built-in bias of tree-ring chronologies, which must rely on trees from Northern Hemisphere regions outside the tropics, because those are the trees with well-defined annual growth rings.

The hockey stick graph was first put together in 1998. The politics soon got going. That year turned out to be the warmest in the instrumental record. So it wasn't much of a stretch to argue that the hockey stick revealed 1998 to be the warmest year in the warmest century of the past millennium. That got headlines. And brought trouble—not least for the voluble, self-confident, and likable collator of the hockey stick data, Mike Mann. Even though the IPCC published other data sets showing much the same, Mann was accused of concocting a spurious case that late-twentieth-century warming was exceptional and therefore, presumably, a result of man-made pollution.

It probably didn't help that at the time, Mann was based at the University of Virginia, home of the biggest voice among the climate skeptics: Pat Michaels. Soon Mann was fraud-of-the-month on the Web sites of the climate skeptics. But the criticism went beyond the normal community of climate skeptics: some serious climate researchers expressed misgivings about Mann's methods.

When I finally met Mann, he had moved from Virginia to Penn State University, where he is now director of the Earth Science Systems Center. But the flak had followed. Some was fair; some was unfair; some was deployed as political hand grenades; some was just a part of the normal adversarial flow of scientific debate; and some was just plain personal —like Wally Broecker's bad-mouthing of Mann, quoted at the start of Chapter 23. Mann was even damned in Washington, where Senator James Inhofe of Oklahoma accused him of playing fast and loose with the data, and Representative Joe Barton of Texas summoned him to provide his committee with voluminous details about working procedures and funding. Some called it a McCarthyite vendetta. But Mann seemed up for it, dismissing Inhofe as "the single largest Senate recipient of oil industry money."

I will now entertain some of the criticisms that have rained on Mann, because they matter. But it is worth saying first that nothing I have heard impugns Mann's scientific integrity, credentials, or motives. He is just braver than some, and more willing to have his debates in public—and to fight back when the brickbats start flying. (You can read him in action on the Web site he started with scientific colleagues at www.realclimate.org.) Some researchers have suffered real personal and psychological damage from attacks by skeptics. I hope that won't happen to Mann. I wish more scientists were like him.

First, does the hockey stick fairly represent the temperature record? Does Mann's take-home conclusion, that the last century warmed faster and further than any other in the past thousand years, stand up to scrutiny? The short answer is yes—but only just.

The world of proxy data trends is a statistical minefield. This is partly because the physical material that shows past climate loses detail with time. Tree rings, for instance, get smaller as the tree gets older, so annual and even decadal detail gets lost. "You lose roughly 40 percent of the amplitude of changes," says the tree ring specialist Gordon Jacoby, of Lamont-Doherty. But it goes far beyond that. To make any sense, analysis of a single data set—for instance, from the tree rings in a forest—involves smoothing out the data from individual trees to reveal a "signal" behind the "noise" of short-term and random change. The kind of analysis pioneered by Mann, in which a series of different data sets are merged, involves further sorting and aggregating these independently derived signals, and smoothing the result. And Mann's work involves a further stage: meshing that proxy synthesis with the current instrumental record.

Some, including Jacoby, complain that by combining smoothed-out proxy data from past centuries with the recent instrumental record, which preserves many more short-term trends, Mann created a false impression of anomalous recent change. "You just can't do that if you are losing so much of the amplitude of change in the rest of the data," Jacoby told me. Mann argues the contrary—that in fact he was one of the first analysts in the field to include error bars on his graph. "The error bars represent how much variance is lost due to the smoothing," he says.

But the accusation that he has somehow fixed the data analysis continues to dog him. The most persistent line of criticism, and the one most widely championed by anti-IPCC lobbyists, came from two Canadians: Stephen McIntyre, a mathematician and oil industry consultant, and Ross McKitrick, an economist at the University of Guelph. They claimed to have found a fundamental flaw in Mann's statistical methodology that biased the temperature reconstruction toward producing the hockey stick shape.

The argument is a technical one that hangs on how Mann used well-established mathematical techniques for classifying data called principal component analysis. McIntyre and McKitrick claimed that Mann's method had the effect of damping down unwanted natural variability, straightening the shaft of the hockey stick and accentuating twentieth-century warming. Mann agrees there was some truth in this charge. He analyzed the data in terms of their divergence from twentieth-century levels, and this had the inevitable effect of giving greater significance to data showing the biggest differences from that period.

But the critical charge was that he had somehow created the hockey stick out of nothing—"mining" the data for hockey-stick-shaped trends, as his critics put it. McIntyre and McKitrick produced their own analysis, showing an apparent rise in temperatures in the fifteenth century, which, they claimed, may have been as warm as the twentieth century. The shaft of the hockey stick had a big kink in it. When it was published, in 2005, this analysis was hailed by some as a refutation of Mann's study.

But while Mann was open to attack, so were McIntyre and McKitrick. Would their "refutation" of Mann stand up to critical attention? During 2005, three different research groups concluded that Mann's findings bear scrutiny much better than do those of his critics. They had bent the statistics more than he had, arbitrarily leaving out certain sets of data to reach their conclusion. Remove all the biases, and the real data looked more like Mann's—a conclusion underlined in early 2006, when Keith Briffa, a respected British tree-ring analyst at the University of East Anglia, published the most complete analysis to date, showing the twentieth century to have been the warmest era for at least the past 1,200 years. Briffa's take was confirmed in June 2006 by the U.S. National Academy of Sciences, which, in a long-awaited review of the hockey-stick debate, endorsed Mann's work. The analysts expressed a "high degree of confidence" that the second half of the twentieth century was warmer than any other period in the previous four centuries. But they said that although many places were clearly warmer now than at any other time since 900, there was simply not enough data to be quite so sure about the period before 1600.

If the key to successful science is producing findings that can be replicated by other groups using different methodologies, then Mann is on a winning streak. Upward of a dozen studies, using both different collections of proxy data and different analytical techniques, have now produced graphs similar to Mann's original hockey stick. Not identical, for sure, but with the same basic features of unremarkable variability for 900 years followed by a sharp upturn in temperatures in the final decades.

The one unexplained factor is that most of these studies show paltry evidence for the medieval warm period and the little ice age. But an answer to that conundrum now seems at hand. There is growing agreement that the most substantial evidence for the existence of both a medieval warm period and a little ice age comes from the northern latitudes. "What we know about the cold in the little-ice-age era is primarily a European and North Atlantic phenomenon," says Keith Briffa. Most interesting, there is growing evidence from a range of new proxy data that other parts of the world were seeing climate trends opposite to those going on in Europe. The tropical Pacific appears to have cooled during the medieval warm period and warmed during the little ice age. One ice core from the Antarctic shows temperatures during the medieval warm period that were 5°F colder than those in the little ice age. Under the circumstances, says Mann, it is not surprising that his more global assessment of temperatures does not spot much difference during these earlier climatic shifts. They undoubtedly had major influences on regional climates, but the cumulative effect on global temperature was small.

It is no part of this book's case that climate didn't change in the past. Parts of the world clearly saw substantial warming and cooling during the medieval warm period and the little ice age. Other parts saw other changes. In the American West, there were huge, century-long droughts during the medieval warm period. Even Broecker, who holds that the little ice age was global, admits that the evidence of a global medieval warm period is "spotty and circumstantial." But there is a good case for saying that over the millennium until the mid-twentieth century, most climate change concerned the redistribution of heat and moisture around the globe rather than big changes in overall heating. Only recently has there been a major additional "forcing," caused by the introduction of hundreds of billions of tons of greenhouse gases into the atmosphere. Recent warming may be the first global warming since the closure of the ice age itself.

The argument over the hockey stick is an interesting sideshow in the debate about climate change. But it remains a sideshow. Right now, it matters little for the planet as a whole whether the medieval warm period was or was not warmer than temperatures today—or, indeed, whether it was a warm period at all. The subtext of the climate skeptics' assault on Mann's hockey stick has always been that if the current warming is shown not to be unique, then somehow the case that man-made global warming is happening evaporates. But this is a spurious argument. Briffa is not alone in arguing precisely the opposite. If it was indeed very warm globally in the medieval warm period, that is truly worrying, he says. "Greater long-term [natural] climatic variability implies a greater sensitivity of climate to forcing, whether from the sun or greenhouse gases. So greater past climate variations imply greater future climate change."

34

HURRICANE SEASON

Raising the storm cones after Katrina

Corky Perret lost everything when Hurricane Katrina hit. His house on the beachfront out on Highway 90 between Gulfport and Biloxi, Mississippi, was reduced to matchwood by 130-mile-an-hour winds, and sucked away by a 30-foot storm surge that washed up the beach and over the highway. "Nothing is left; it was totally destroyed," he told me weeks later. Out in the Gulf of Mexico, barrier islands that once provided protection against storms had also succumbed. Perret didn't know if hurricanes would be worse in the future, but without the islands, the effects would probably be worse anyway.

The houses along the section of Highway 90 where Perret lived, along with the hotels and resorts, had been built mostly between the 1970s and the 1990s, a period of quiet in the Gulf when there were few hurricanes. Hearing reports that no letup is likely anytime soon, some of his neighbors were going for good. They could see only more hurricanes and more havoc. They were off to Jackson or Dallas or Memphis, or anywhere inland. But when we spoke in late 2005, Perret still had his job as director of marine fisheries for the state of Mississippi, and was unsure what to do. He wanted to stay and rebuild, but was that wise?

The year 2005 had been an extraordinary one in the Atlantic. There were so many tropical storms that for the first time meteorologists ran out of names for them. Wilma became the most powerful Atlantic storm ever recorded. Katrina brought an entire U.S. city to its knees. It was the second hurricane year in a row to be described by meteorologists as "exceptional" and "unprecedented," and it came after a decade of rising hurricane activity that stretched the bounds of what had previously been regarded as natural. So what was going on? Are hurricanes becoming more destructive as global warming kicks in? Is there worse to come? The answer matters not just to the people in the firing line around the Gulf of Mexico and the Caribbean, or across the tropics in the Indian Ocean and the Pacific: if there's more severe disruption to oil production in the Gulf, or super-typhoons hit economic powerhouses like Shanghai or Tokyo at full force, we'll all feel the impact.

Until 2005, most of the world's leading hurricane experts were san-guine. The upsurge in the number of hurricanes in the North Atlantic in the previous decade had been just part of a normal cycle. Hurricanes had been strong before, from the 1940s through the 1960s. Climate models suggested that even a doubling of carbon dioxide levels in the atmosphere would increase hurricane intensity by only 10 percent or so. But that year the consensus was shattered. A flurry of papers claimed that hurricanes had grown more intense during the past thirty-year surge in global tempera-tures. Not more frequent, but more intense, with stronger winds, longer durations, more unrelenting rains, and even less predictable tracks. The trend was apparent in all the world's oceans, they said. From New Orleans to Tokyo, nobody was immune.

One of the authors, Kerry Emanuel, of the Massachusetts Institute of Technology, said: "My results suggest that future warming may lead to an upward trend in tropical cyclone destructive potential and—taking into account an increasing coastal population—a substantial increase in hurricane-related losses in the 21st century." Coming just weeks after the destruction of New Orleans, that sounded like a clear message to Corky Perret and the people of the Gulf Coast. No point in rebuilding, because the next superhurricane could be just around the corner. But the claims produced a schism among the high priests of hurricane forecasting. Many, like the veteran forecaster William Gray, of Colorado State University, said that they saw no upward trend and no human fingerprint. They accused the authors of the latest papers of bias and worse. So who was right?

Hurricanes are an established part of the climate system. There have al-ways been hurricanes. They start off as clusters of thunderstorms that form as warm, humid air rises from the surface of the tropical ocean. As the air rises, the water vapor condenses, releasing energy that heats the air and makes it rise even higher. If enough storm clouds gather in close proximity, they can form what Emanuel calls a "pillar" of humid air, extending from the ocean surface for several miles into the troposphere. The low pressure at the base of the pillar sucks in more air, which picks up energy in the form of water vapor as it flows inward, and releases it as it rises. This lowers the pressure still further.

Meanwhile, the rotation of Earth, acting on the inward-flowing air, makes the pillar spin. If conditions are favorable, a tropical storm can rapidly turn into a hurricane as wind speeds pick up. Its power is staggering: Chris Landsea, of the National Oceanic and Atmospheric Administration, in Miami, has calculated that an average hurricane can release in a day as much energy as a million Hiroshima bombs. Luckily for all concerned, only a tiny fraction of this energy is converted into winds.

Worldwide there are about eighty-five tropical cyclones each year, of which about sixty reach hurricane force. That figure has been fairly stable for as long as people have been counting hurricanes. But the distribution of the hurricanes varies a great deal from year to year. In 2005, for example, the Atlantic was battered but the Pacific was relatively peaceful. On the face of it, global warming is likely to make things worse. The initial pillar of humid air forms only when the temperature of the sea surface exceeds 78°F. As the world's oceans warm, ever-larger areas of ocean exceed the threshold. There has been an average ocean warming in the tropics of 0.5 degrees already.

What is more, every degree above the threshold seems to encourage stronger hurricanes. When Katrina went from a category 1 to a category 5 hurricane back in August 2005, the surface of the Gulf of Mexico was around 86°F, which, so far as anyone knows, was a record. Whether or not climate change can be blamed for the record sea temperatures (and most would guess that it can), those temperatures certainly helped Katrina strengthen as it slipped across the Gulf from Florida toward the Louisiana coast.

This simple link between sea surface temperatures and hurricane formation and strength has encouraged the view that a warmer world will inevitably lead to more hurricanes, stronger hurricanes, and the formation of hurricanes in places formerly outside their range. But the world is not that simple, says William Gray. What actually drives the updrafts that create the storm clouds, he says, is not the absolute temperature at the sea's surface but the difference in temperature at the top of the storm. Climate models suggest that global warming will raise air temperatures aloft. So, if he is right, while the current sea surface temperatures necessary to create hurricanes may be 78°F or more, it could in future rise to 82° or more. In the final analysis, Gray argues, the hurricane-generating potential of the tropics may remain largely unchanged.

There are other limitations on hurricane formation. However hot the oceans get, air cannot rise everywhere. It has to fall in some places, too, whatever the ocean temperature. And many incipient hurricanes are defused by horizontal winds that lop off their tops. Climate models suggest that global warming will increase wind speeds at levels where they would disrupt hurricanes. Other disruptions include dust, which often blows across the Atlantic during dry years in the Sahara.

But some trends will make big storms more likely. Most tropical storms fizzle out because they lose contact with their fuel—the heat of warm ocean waters. This happens most obviously when a hurricane passes over land, but it also happens at sea. As the storm grows, its waves stir up the ocean, mixing the warm surface water with the generally cooler water beneath. The surface water cools, and that can be the end. In practice, a hurricane can grow only if the warmth extends for tens of yards or more below the surface. But with every year that passes, warm water is penetrating ever deeper into the world's oceans. That is clearly tied to global warming. And it is setting up ideal conditions for more violent thunderstorms. Katrina is again an object lesson here. It continued to strengthen as it headed toward New Orleans, because it moved over water in the Gulf of Mexico that was very warm, not just at the surface but to a depth of more than 300 feet.

The past decade in the North Atlantic has seen a string of records broken. The period from 1995 to 1998 experienced more Atlantic hurricanes than had ever before occurred in such a short time—a record broken only in 2004 and 2005. The 1998 season was the first in a 100-year record when, on September 25, four hurricanes were on weather charts of the North Atlantic at one time. And not long afterward came Hurricane Mitch, the most destructive storm in the Western Hemisphere for 200 years. Feeding on exceptionally warm waters in the Caribbean, it ripped through Central America in the final days of October 1998, its torrential rains bringing havoc to Honduras and Nicaragua and killing some 10,000 people in landslides and floods.

The Atlantic is also generating hurricanes in places where they have never been seen before. In March 2004, the first known hurricane in the South Atlantic formed, striking southern Brazil. That the hurricane, later named Catarina, even formed was startling enough. What caused the greatest shock was that it developed very close to a zone of ocean pinpointed a few years before by Britain's Hadley Centre modelers as a likely new focus for hurricane formation in a warmer greenhouse world. But they had predicted that the waters there wouldn't be up to the task till 2070. Many saw Catarina as a further sign that global warming was making its presence felt in the hurricane world rather ahead of schedule.

The billion-dollar question (literally so for insurance companies) is whether there is now a discernible climate change component at work in the frequency and intensity of hurricanes. Kerry Emanuel, for one, argues that whatever the natural variability, the "large upswing" in hurricanes in the North Atlantic in the past decade is "unprecedented, and probably reflects the effect of global warming." Jim Hansen weighed in at the end of 2005, insisting that climate change was the cause of a warmer tropical Atlantic and that "the contention that hurricane formation has nothing to do with global warming seems irrational and untenable."

The matter of North Atlantic hurricane trends is likely to be debated for many years yet. The "signal" of climate change will be difficult to disentangle from the "noise" of natural variability. But while it is easy to become obsessed with hurricanes in the North Atlantic, they amount to only around a tenth of the global total—and a rather smaller proportion of those that make landfall in a typical year. The biggest source of hurricanes is, and is likely to remain, in the western Pacific, where they terrorize vulnerable and densely populated nations like the Philippines, Vietnam, and China. So it is the global picture that both matters most and is most likely to resolve the issue of the impact of climate change.

Several research groups have been scouring records of past hurricanes worldwide to see if there is any evidence of a trend as the world has warmed. Emanuel has concluded that, on average, storms are lasting 60 percent longer and generating wind speeds 15 percent higher than they did back in the 1950s. The damage done by a hurricane is proportional not to the wind speed but to the wind speed cubed. And Emanuel's results suggest that the destructive power of a typical hurricane has increased by an alarming 70 percent. "Global tropical cyclone activity is responding in a rather large way to global warming," he says.

Others are coming to agree. Only weeks after Emanuel's paper appeared, in the autumn of 2005, three other leading hurricane researchers published a similarly alarming conclusion. Peter Webster and Judy Curry, of the Georgia Institute of Technology, and Greg Holland, of NCAR, concluded that while there had been no overall increase in the number of hurricanes worldwide, the frequency of the strongest storms—categories 4 and 5—had almost doubled since the early 1970s. They now made up 35 percent of the total, compared with 20 percent just three decades before. The trend, the researchers said, was global, and they agreed with Emanuel that it was clearly connected to the worldwide rise in sea surface temperatures. That made it extremely unlikely that natural cycles, which are relatively short-term and confined to single ocean basins, were causing the trend. "We can say with confidence that the trends in sea surface temperatures and hurricane intensity are connected to climate change," Curry declared.

William Gray and some other traditional hurricane forecasters have contested the findings, claiming that some of the data, particularly old estimates of wind speed from the Pacific in the 1970s, are flawed. In an increasingly vitriolic exchange, Gray argued that the papers simply could not be true. Emanuel and Webster agree that the data are not as good as they might like. But "Gray has not brought to my attention any difficulties with the data [of] which I was not already aware," Emanuel said, with some irritation. Webster says Gray is "grasping at thin air."

So where does that leave us? There is as yet nothing unique about recent individual hurricanes, though Katrina, Wilma, and Mitch clearly stretch the bounds of what can be regarded as normal. The largest and most powerful hurricane ever recorded, Typhoon Tip, with wind speeds of more than 180 miles per hour, grazed Japan a quarter of a century ago, in 1979. The storm that hit Galveston in 1900 killed 10,000 people, many more than Katrina. Both pale compared with a hurricane in 1970 that may have killed half a million people in what is now Bangladesh.

But even if we don't yet see "superhurricanes," evidence is emerging of a human fingerprint in the rising number of stronger, longer-lasting hurricanes. It is not yet proof of a long-term global trend tied to global warming, but the striking finding from both Emanuel and Webster that there is a consistent, global connection between rising sea surface temperature and rising storm strength is strong evidence of such a link. Whatever the theoretical concerns, for now it seems that, as the climatologist Kevin Trenberth, of the National Oceanic and Atmospheric Administration, puts it: "High sea surface temperatures make for more intense storms." In a paper published in June 2006, Trenberth calculated that about half of the extra warmth in the waters of the tropical North Atlantic in 2005 could be attributed to global warming. This warming, he said, "provides a new background level that increases the risks of future enhancements in hurricane activity."

One puzzling question is how scientists have until now failed to spot the sharply increased destructiveness of modern hurricanes. There is no dispute that, taken together, hurricanes have been doing a lot more damage in recent years. In badly organized countries, such as many in Central America, that has often meant a heavy loss of life. Elsewhere, if evacuation systems work, it has simply meant a huge loss of property. Insurance claims for hurricane disasters have been soaring for some years.

The prevailing view has, until recently, been that the problem is one of bad planning, rising populations, and more people putting themselves in harm's way. The beach resorts along Highway 90 and the large squatter colonies spreading along low-lying coastal land in Asia give some support to that view. But the new data suggest that there is more to it than that. A lot more. And that most of the extra damage is being caused by the storms themselves becoming more intense. The trend seems set to continue.

35

OZONE HOLES IN THE GREENHOUSE

Why millions face radiation threat

Joe Farman is a scientist of the old school. String and sealing wax. Smokes a pipe and drinks real ale. He has the faraway look in his eyes that you often see in men who have spent any length of time in Antarctica. He is retired now from the British Antarctic Survey, where he spent virtually his entire working life in a worthy though less than exalted capacity. Or he did until 1985, when he wrote one of the decade's most quoted research papers. He is the man who discovered the ozone hole over Antarctica. And the way it happened—or, rather, almost didn't happen—is revealing.

A quarter of a century ago, Farman was in charge of the BAS's Dobson meter, which for many years had been pointing up into the sky measuring the depth of the ozone layer in the stratosphere from the BAS's base at Halley Bay, on an ice shelf off West Antarctica. For several years his bosses had been trying to halt the observations and bring the old instrument home. After all, they pointed out, nothing interesting had happened for years, and satellites orbiting Earth were by then measuring ozone levels routinely. Ground-based observations were deemed superfluous.

But Farman resisted, and in 1982, he noticed a series of unusual and abrupt fluctuations in the ozone readings, just after the sun reappeared following the long polar night. It happened again the following year.

"I asked the Americans if they had seen anything similar from their satellites," he told me later. "They said they hadn't. So I assumed that my old machine was on the blink." But he was intrigued enough not to leave it at that. He found another Dobson meter back in Cambridge, and took it south in 1984 to check the readings. It recorded the same thing—only more so. Farman's data were by now unambiguous. He was seeing a deep hole opening in the ozone layer over the base. It lasted for several weeks before closing again. "We were sure then that something dramatic was happening," Farman said. In places, more than 90 percent of the ozone was disappearing in what appeared to be runaway reactions taking place in just a few days.

The ozone layer protects Earth's surface from dangerous ultraviolet radiation from the sun. Without this filter, there would be epidemics of skin cancers, cataracts, and many other diseases, as well as damage to vital ecosystems. Life on Earth has evolved to live under its protection, and would find things much harder without it.

For more than a decade, scientists had been concerned about the ozone layer, fearing that man-made chemicals such CFCs in aerosols might cause it to thin. But nobody had thought of a hole forming. Least of all over Antarctica, which was as far from the source of any ozone-destroying chemicals as you could get. And certainly not in a runaway reaction over just a few days. Earth was simply not supposed to work that way.

Farman bit his pipe and got to work. No more checking with NASA. He had his data and was intent on an urgent publication in the scientific press. Perhaps he sensed it was his moment of fame. He was certainly scared by what he had found—scared enough to miss all the office parties in Cambridge in 1984 to finish his paper titled "Large Losses of Total Ozone in Antarctica." He posted it to *Nature* on Christmas Eve.

The editors didn't quite share Farman's sense of urgency. It took them three months to accept his paper, and another two months to publish it. When the paper finally appeared, NASA scientists were confused. They still had no inkling of anything amiss over Antarctica. But they could hardly ignore the findings of two Dobson meters, however ancient. They re-examined the raw data from their satellite instruments and were shocked to find that their satellites had seen the ozone hole forming and growing over Antarctica all along, even before Farman had spotted it. But the computers on the ground that were analyzing the streams of data had been programmed to throw out any wildly abnormal readings. And the data showing the ozone hole had certainly fitted that category. The episode, as Farman was not slow to point out, was a salutary lesson for high-tech science. It was also a triumph for the string-and-sealing-wax school, and for the dogged collection of seemingly boring and useless data about the environment.

Paul Crutzen—who had unraveled much of the complex chemistry of the ozone layer—swiftly tied Farman's findings to specific chemical reactions involving CFCs that took place only in the uniquely cold air over Antarctica each spring. Below about -130°F, unique clouds form in the stratosphere above Antarctica. These are called polar stratospheric clouds. It turned out that the runaway reactions happened only on the surface of the frozen particles in these clouds. The reactions required both the cold to create the clouds and solar energy to fuel them. And there was a window of a few weeks when both were supplied—after the sun had risen, but before the air warmed enough to destroy the clouds. After that, the air warmed and the ozone recovered, though the repair job took some months.

Farman's discovery and Crutzen's analysis finally pushed the world into taking tough action against ozone-eating chemicals. The Montreal Protocol was signed in 1987. Slowly, very slowly, the amount of CFCs and other ozone-eaters in the stratosphere is declining. And the Antarctic ozone layer is equally slowly starting to heal, though it could be a century before it is fully repaired, even if every promise made by government negotiators is met. But it had been a close call.

And things could have been a lot worse. "Looking back, we were extremely lucky that industrialists chose chlorine compounds, rather than the very similar bromine compounds, to put in spray cans and refrigerators early in the last century," says Crutzen. Why so? Bromine compounds make refrigerants that are at least as effective as their chlorine equivalents. But atom for atom, bromine is about a hundred times better than chlorine at destroying ozone. Pure luck determined that Thomas Midgley, the American chemist who developed CFCs, did not opt for their bromine equivalent. "It is a nightmarish thought," says Crutzen, "but if he had chosen bromine, we would have had something far worse than an ozone hole over Antarctica. We would have been faced with a catastrophic ozone hole, everywhere and at all seasons during the 1970s, before we knew a thing about what was going on."

The world has been very lucky. Or has been lucky so far. The same combination of low temperatures and accumulating gases that combined so devastatingly over Antarctica can also occur over the Arctic in some years. The conditions are not quite so favorable for ozone destruction, because the atmosphere is not quite so stable and the extremely cold temperatures occur less frequently. But there have been some near misses.

One occurred in January 2005. Anne Hormes, who runs the German research station at Ny-Alesund, in Svalbard, told me the story when I visited there a few months later. Temperatures in the lower stratosphere above Svalbard had for a few days fallen to -144°F, fully 14 degrees below the threshold necessary for the formation of polar stratospheric clouds, and extremely low even by the standards of Antarctica. "We feared that a real, big ozone hole would form," she said. "And if the temperature had stayed that cold for a few more weeks, till the sun came up to drive the chemical reactions, we would certainly have seen one." It would have been the Arctic's first full-fledged ozone hole, and in all probability a major world environment story.

Her concern is shared. The ozone expert Drew Shindell, of the Goddard Institute for Space Studies, says: "Overall winter temperatures are going down in the Arctic stratosphere—2005 was very cold. But actual ozone loss is very time-critical. So far, we have been lucky." But he doubts that our luck will hold. How so? Why are the risks of an ozone hole still growing, even though the chemicals that cause it are now in decline in the stratosphere?

The problem is this. In the lower atmosphere, greenhouse gases trap heat. But in the stratosphere, they have the opposite effect, causing an increase in the amount of heat that escapes to space from that zone of the atmosphere. This is happening worldwide, but some of the most intense stratospheric cooling is over areas with the greatest warming at the surface. Like the Arctic, where the air increasingly resembles the air high above Antarctica.

There is another risk factor, too. The warmer troposphere, with stronger convection currents taking thunderstorm clouds right up to the boundary with the stratosphere, may be injecting more water vapor into the stratosphere. As far as we know, the stratosphere has always been very dry in the past. So extra water vapor is potentially a big change. And more water vapor will make more likely the formation of the polar stratospheric clouds within which ozone destruction takes place. "If it gets a lot wetter, that will make ozone depletion much worse," says Shindell. There is some evidence that that is happening, though data are scarce. "Water vapor levels in parts of the lower stratosphere have doubled in the past sixty years," he says.

No hole formed in the Arctic ozone layer in 2005, because the sun did not rise when the air was at its coldest. But the spring of 2005 nonetheless saw the largest Arctic ozone loss in forty years of record-keeping. More than a third of the ozone disappeared, and losses reached 70 percent in some places. Air masses with reduced ozone levels spread south across Scandinavia and Britain, and even as far south as Italy for a few days. One year soon, the sun will rise when temperatures are still cold enough for major runaway ozone destruction. And when it does, millions of people may be living beneath. This will be another unexpected consequence of global warming.

VIII

INEVITABLE SURPRISES

36

THE DANCE

The poles or the tropics? Who leads in the climatic dance?

As we have seen, researchers into the global history of climate, especially in the U.S., divide into two camps. One believes that the key drivers for past, and therefore probably future, climate change lie in the polar regions, especially the far North Atlantic. The other believes that the real action happens in the tropics.

The most outspoken advocate for the polar school is Wally Broecker, of Lamont-Doherty. As described in Chapter 23, he is the man behind the idea of the ocean conveyor, which begins in the far North Atlantic and which, he argues, is the great climatic amplifier. It has, he says, a simple on-off switch. It pushed the world into and out of ice ages; it modulates the effects of Bond's solar pulse, including its most recent manifestations in the medieval warm period and the little ice age; and it could be a big player in directing the consequences of global warming. Around Broecker is a whole school of researchers who have spent their careers investigating the dramatic climate events of the North Atlantic region, as recorded in the ice cores of Greenland.

The rival, tropical school has often looked to two characters. One, just down the corridor from Broecker at Lamont-Doherty, is Mark Cane, a leading modeler of El Niño, the biggest climate fluctuation in the tropics. The other is Lonnie Thompson, the man who decided thirty years ago to stop investigating polar ice cores and switch instead to drilling tropical glaciers. They argue that Broecker's ocean conveyor is at best a sideshow, relevant to the North Atlantic and the countries that border it, but not the great global amplifier it is claimed to be. For them, the important climatic levers must be in the planetary heat and hydrological engines around

Earth's girth. The debate between the two schools has, at various stages, become quite personal. "It all came from one man: Wally Broecker," says Cane. "You were for him or against him. And I found myself against."

The polar people deploy their polar ice core data to show that climate change has been more dramatic and sudden in the far North, so that must be the cockpit of climate change. This is where the Gulf Stream turns turtle and drives the ocean conveyor; this is where ice melting and changes in freshwater flow can freeze the ocean virtually overnight and send temperatures tumbling by tens of degrees; this, above all, is where the great ice sheets of the ice ages formed and died. They have a point. There can be little doubt about the importance of ice formation to the ice ages. Virtually the whole world cooled then, and two thirds of that cooling was caused by the feedback of growing ice sheets and their ability to reflect solar radiation back into space. And nothing except a huge rush of meltwater from the receding ice caps could have plunged the world into the Younger Dryas, 12,800 years ago.

But that doesn't mean that the Arctic tells the whole story. What pulled the world out of the Younger Dryas, for instance—an event that happened even faster than its onset? And while big climate change during and at the close of the ice ages does seem to be associated with polar events, the evidence concerning climate change since is far less secure. Thompson argues that most of the global climatic shudders of the Holocene, such as events 5,500 and 4,200 years ago, must have been tropical in origin: "In climate models, you can only make such things happen in both the Northern and Southern Hemispheres by forcing events from the tropics, and I am convinced that is what is happening."

Hockey-stick author Mike Mann, though not a fully paid-up member of the tropical school, says: "I increasingly think that the tropical Pacific is the key player. When you see La Niña dominating the medieval warm period and El Niño taking hold in the little ice age, it begins to look like the tropics, rather than the North Atlantic, rule." The argument is that heat flows from the tropics are the true intermediaries between Bond's solar pulse and temperature fluctuations in the North Atlantic.

The tropical school also accuses the polar fraternity of being blinkered about what constitutes climate change. Besides overly focusing on events in North America and Europe, it stands accused of being overly concerned with temperature. In the tropics, the hydrological cycle matters more than the temperature. Megadroughts are as damaging as little ice ages, and the rains, rather than extra warmth, bring plenty. Witness the drying of the Sahara 5,500 years ago, and the importance of the vagaries of the Asian monsoon.

The tropical school doesn't stop there. Its adherents argue that many of the big climatic events in the Northern polar regions have their origins in the tropics. The tropics, by delivering warm water into the North Atlantic, are just as capable of flipping the switch of the ocean conveyor as is ice formation in the far North Atlantic. And if there is a tropical equivalent of Broecker's switch in the North Atlantic, they say, it is probably the warm water pool around Indonesia—an area they often call "the firebox." This is the greatest store and distribution point for heat on the surface of the planet, with a known propensity for threshold changes via the El Niño system. It is also the biggest generator of water vapor for the atmosphere, which is both a potent greenhouse gas and a driver of weather systems.

If this region can trigger short-term El Niños that warm the whole planet, and La Niñas that cool it again, then might it not also trigger long-term climate changes? Might not events here have been important in turning a minor orbital wobble into the waxing and waning of the ice ages? The waning, certainly. For cores of ocean sediment recently taken from the tropical Pacific suggest that temperatures started to rise there a thousand or more years before the Northern ice sheets began to shrink.

But after some years of standoff, many protagonists in this debate are now seeking common ground. Not Broecker, of course. But Richard Alley, a polar man but also a fan of Thompson's, now thinks that the location of the climate system driver's seat may change with time. It is easy to imagine the power of ice and meltwater to hijack the world's climate during the glaciations, when a third of the Northern Hemisphere was covered with ice. But with less ice around in the interglacials, he concedes, the argument is less persuasive. And, with characteristic pithiness, he admits to past regional bias. "Suppose the North Atlantic circulation did shut down. Sure, Europe would care. They might have a midseason break in football in Britain. Manchester United wouldn't be playing on Boxing Day. But in the Great Plains of the U.S. and in the Pacific Ocean, would it be so important?"

Meanwhile, on the tropical side, Cane admits: "I am less absolutist than I used to be." He agrees that his great enthusiasms, El Niño and the tropical Pacific, might not be behind everything. He still believes that the role of the ocean conveyor is hopelessly hyped, but he concedes the possible importance of the "sink or freeze" switch for sea ice in the North Atlantic. The divide between the polar and tropical schools is "a slightly false separation," says Peter deMenocal, of Lamont-Doherty. "You cannot at the end of the day change one bit without changing the other. They are all part of the same pattern, whether leading or following." Earth functions as an integrated system, not as a series of discrete levers.

That view seems to be confirmed by Steve Goldstein, of Columbia University, who has used analysis of a rare earth called neodymium, which has different isotopic ratios in different oceans, to reconstruct the order of events at the starts and ends of the ice ages. He argues that orbital changes, as expected, lead events. But the first feedback to respond is the ice-albedo feedback. It caused an initial cooling at the start of the last ice age that was most pronounced in the far North. Prompted by that initial cooling, the chemistry and biology of the oceans started to change, removing carbon dioxide from the atmosphere and accentuating the cooling further. Only then, some thousands of years later, did the ocean conveyor start to shut down. "The conveyor follows; it does not lead," he says. If his analysis is confirmed, it will be a blow to Broecker, but it will also confirm that both the tropics and the polar regions were deeply implicated in the elaborate dance that took the world into and out of the ice ages.

Paul Crutzen has been in the forefront of research in both spheres, helping crack the mysteries of the Antarctic ozone layer while making a strong case for the dynamic properties of the tropical heat engine. "Big planetary changes happen in both the tropics and the very high latitudes," he says. "The tropics are where the high temperatures drive a lot of the chemistry and dynamics of the atmosphere. And the polar regions are the homes of the big natural feedbacks that could accelerate climate change: things like melting ice and permafrost and alterations to ocean currents." That is probably as good a compromise statement as can be found right now. At the end of the day, the system is bigger than the individual parts.

37

NEW HORIZONS

Feedbacks from the stratosphere

Is that the end of the story? I don't think so. Constantly, in writing this book, I have been struck by how little we know about the way Earth's climate and its attendant systems, feedbacks, and oscillations function. This story contains some heroic guesses, some brilliant intuition, and, no doubt, occasionally some dreadful howlers—because that is where the science currently lies. More questions than answers. Beyond the cautious certainties of the IPCC reports, there is a swath of conjectures and scary scenarios. Some criticize the scientists who talk about these possibilities for failing to stick to certainties, and for rocking the IPCC's boat. But I suspect we still need a good deal more of the same, because we may know much less than we think. I think Wally Broecker and his colleagues deserve praise for developing their scenarios about the global conveyor. They have produced a persuasive narrative that has transformed debate. Of course, producing a persuasive story doesn't make it right, but it does generate new research and new ideas that can be tested. It is time someone in the tropical school produced something comparable.

Equally important, there may be other narratives that need developing. Richard Alley must be right that there are more "inevitable surprises" out there—outcomes that nobody has yet thought of, let alone tested. One area where unconsidered triggers for global climate change may lie is in and around Antarctica. While sinking cores into Antarctica as well as Greenland, the polar school has yet to devote much attention to generating theories about events in the South Atlantic. This may be a mistake. Much of the action in Earth-system science in the next few years will happen there, I am sure. Any place capable of producing something as remarkable as the ozone hole in the stratosphere is surely capable of storing up other surprises.

One new idea emerging from the battle between the polar and tropical schools is that the real driver of climate change up to and including the ice ages may actually lie in the far South. During ice ages, the theory goes, the ocean conveyor did not so much shut down as start getting its new deep water from the Antarctic rather than the Arctic. A certain amount of deep water has always formed around Antarctica, though in recent times it has played second fiddle to the North Atlantic. But, as the ice sheets grew across the Arctic and the chimneys in the North Atlantic shut down, the zone of deepwater formation in the Southern Ocean seems to have strengthened and may have taken charge of the conveyor.

Some go further and say that there must be a "bipolar seesaw," in which warming in the Southern Hemisphere is tied to cooling in the North and vice versa. That would certainly make sense of some of the Antarctic ice cores that show warming while the North was cooling. The question then is: Which pole leads? Does the North Atlantic end of the system shut down, closing off the Gulf Stream's northward flow of warm water and leaving more heat in the South Atlantic? Or does some switch in the South trigger the shutdown of the Gulf Stream and leave the Northern Hemisphere out in the cold, with the North Atlantic freezing over?

The idea that the South may lead in this particular dance gained ground late in 2005, when results were published from new ice cores in Antarctica. A European group found that the tightest "coupling" between temperature and carbon dioxide levels in the atmosphere is to be found in Antarctic cores, rather than their Greenland equivalents. "The way I see things is that the tropics and Antarctica are in phase and lead the North Atlantic," says Peter deMenocal, of Lamont-Doherty. "Even though we may see the largest events in the North Atlantic, they are often responding, not leading." By this reading, the onset of the Northern glaciation may have its origins in the Southern Hemisphere.

This apparently obscure debate could matter a great deal in the twenty-first century. Right now, the world has become worried that melting ice in the Arctic could freshen the far North Atlantic and shut down the Gulf Stream. This is a real fear. But maybe, while we are researching that possibility, we are ignoring the risk that large stores of freshwater in the Antarctic might break out and disrupt deepwater formation there. Arguably, the risks are far greater in the South, where, besides the potential breakout of ice from Pine Island Bay, recent radar mapping studies have revealed a large number of lakes of liquid water beneath the ice sheets of Antarctica. They might set off a cascade of freshwater into the Southern Ocean, similar in scale to the emptying of Lake Agassiz. Yet nobody, so far as I am aware, has studied what the effects of such a breakout might be for deepwater formation and the Southern arm of the ocean conveyor.

Or, rather than shutting down deepwater formation in Antarctica, might we be about to trigger a switch in the bipolar seesaw, so that deepwater formation in the South takes over from that in the far North? Could that switch be flipped in the South, rather than in the North? And if so, how? And what might happen? It would certainly lead to the Southern Hemisphere's hanging on to very large amounts of heat that currently head north on the Gulf Stream. The Southern Ocean might warm dramatically while the North Atlantic froze. And if the Southern Ocean were to warm substantially, says Will Steffen, the former head of the International Geosphere-Biosphere Programme, "it could result in the surging, melting, and collapse of the West Antarctic ice sheet." Ouch.

If anybody doubts that plenty of new surprises are waiting to be discovered, then the work by Drew Shindell, of the Goddard Institute for Space Studies (GISS), should offer food for thought. His story starts with an apparent success for climate modelers. Since the days of Arrhenius, most climate models have predicted that global warming will be greatest at high latitudes, where known feedbacks like ice-albedo are most pronounced. So rises in temperatures of up to 5°F over parts of the Arctic and the Antarctic Peninsula in recent decades have often been taken as the first proof of man-made climate change.

But there has been a persistent and troubling counterargument. The warming in the polar regions appears to be linked to two natural climatic fluctuations, one in the North and one in the South. In the North, the fluctuation is known as the Arctic Oscillation, an extension of the better-known North Atlantic Oscillation. It is the second largest climate cycle

on Earth, after El Niño. The oscillation itself, as measured by meteorologists, is a change in relative air pressure, but its main impact is to
strengthen or weaken the prevailing westerly winds that circle the Arctic.
Like El Niño, the Arctic Oscillation flips between two modes. In its positive mode, air pressure differences between the polar and extrapolar regions
are strong, and winds strengthen. Especially in winter, the winds take heat
from the warm oceans and heat the land. So, during a positive phase of the
Arctic Oscillation, northern Europe, Svalbard, Siberia, the Atlantic coast
of North America, and Alaska all warm strongly. Likewise, when the oscillation is in its negative phase, the winds drop and the land cools.

The strength of this effect depends on the warmth of the oceans, and
in particular on the Gulf Stream and the health of the ocean conveyor. But
for most of the past thirty-five years, the Arctic Oscillation has been in
a strongly positive mode, helping sustain a long period of warming. Modeling studies suggest that at least half of the warming in parts of the
Northern Hemisphere is directly due to its influence, leaving global
warming itself apparently a bit player. Except that there is growing evidence that global warming is driving the Arctic Oscillation, too. And it
does so from a surprising direction.

Enter Shindell. He likes to occupy the unpopular boundaries between
scientific disciplines. His particular interest is the little-studied relationship between the stratosphere, the home of the ozone layer, and the troposphere, where our weather happens. He studies this with the aid of the
GISS climate model, one of the few that can fully include the stratosphere
in its calculations. Most models show little relationship between global
warming and the Arctic Oscillation. The GISS model is the same when the
stratosphere is not included. But Shindell discovered that when the stratosphere is hooked up, the result is a huge intensification of the Arctic Oscillation and the westerly winds around the Arctic. In fact, with current
levels of greenhouse gases, he has reproduced a pattern very similar to the
current unusually strong positive state of the oscillation.

What is going on? One of the problems with climate models is that it
is not always easy to pinpoint exactly which of the elements in the model
is causing the effects that you see in the printout. But here the role of the
stratosphere is clear. And Shindell reckons he has the links in the chain ex-

plained, at least. As greenhouse gases cool the stratosphere, this cooling alters energy distribution within so as to strengthen stratospheric winds. In particular, a wind called the stratospheric jet, which swirls around the Arctic each winter, picks up speed. This wind, in turn, drives the westerly winds beneath, in the troposphere. So they go faster, too. In this way, a stratospheric feedback is amplifying global warming in the Arctic region by pushing the Arctic Oscillation into overdrive and strengthening the winds that warm the land. It is a brilliant, startling, and, until recently, entirely unexpected feedback.

Might the same apply to events in Antarctica? The GISS model suggests so. There, the dominant climatic oscillation is the Southern Hemisphere annular mode, or SAM. Like the Arctic Oscillation, the SAM is a measure of the air pressure difference between polar and nonpolar air that drives westerly winds sweeping around Antarctica. The geography is somewhat different from the Arctic's. The winds whistle around the Southern Ocean and hit land only on the Antarctic Peninsula, which juts out from the Antarctic mainland toward South America.

The climatologist John King has studied the SAM for the British Antarctic Survey. He says that, like the Arctic Oscillation, it has been in overdrive since the mid-1960s, driving stronger westerly winds. And, again like the Arctic Oscillation, it is amplifying warming along its path. The Antarctic Peninsula has seen air temperatures rise by 5°F since the 1960s—the only spot in the Southern Hemisphere to show warming on this scale. The effects include the melting of the peninsula's glaciers and the dramatic collapse of its floating ice shelves, such as the Larsen B. Additionally, by bringing more warm air farther south, the SAM winds are warming the waters that wash around the edges of Antarctica and beneath its ice—helping destabilize the West Antarctic ice sheet.

Here again, Shindell's model suggests that the strengthening of the SAM is the product of a cooling stratosphere and a strengthening of stratospheric jets. There is an important additional element here in the thinning ozone layer, which makes an additional contribution to stratospheric cooling.

All this is alarming evidence of a new positive feedback that intensifies warming in two particularly sensitive regions of the planet, where that extra warming could unleash further dangerous change. Glaciologists say that the Greenland ice sheet could collapse if warming there reaches 5°F. The huge stores of methane beneath the Siberian permafrost and the Barents Sea could be liberated by similar warming. And "the SAM warming now includes parts of the West Antarctic ice sheet, as well as the Antarctic Peninsula," says Shindell's boss, Jim Hansen. "This is a really urgent issue."

The discovery of the stratospheric feedback also helps answer another question that has long bothered climate scientists: Why do variations in solar output that are probably no more than half a watt per 10.8 square feet cause the big climate fluctuations in the North Atlantic identified by Gerard Bond in his analysis of the 1,500-year solar pulse? Conventional climate models without a stratospheric dimension suggest that such a solar fluctuation shouldn't produce temperature changes of more than 0.35°F. But, although the global temperature change may well have been close to that, in parts of Europe and North America the pulses produce changes ten times as great.

Researchers have struggled to find amplifying mechanisms that might have caused that. Sea ice, the ocean conveyor, and tropical flips like El Niño have all been suggested, but none seems up to the task. Shindell says the answer is his stratospheric feedback. The heart of the mechanism this time is ultraviolet radiation. While the total solar radiation reaching Earth's surface during Bond's pulses varies by only a tenth of a percentage point, the amount of ultraviolet radiation reaching Earth changes by as much as 10 percent. Most of the ultraviolet radiation is absorbed by the ozone layer in the stratosphere, so its impact at ground level is small. But the process of absorption causes important changes in energy flows in the stratosphere. These eventually change the stratospheric jets, and with them the Arctic Oscillation in the Northern Hemisphere and the SAM in the South.

Shindell modeled the likely effects of the last reduction of solar radiation at the Maunder Minimum in the depths of Europe's little ice age, 350 years ago. The GISS model without the stratosphere was unmoved by the tiny change in solar radiation. But with the stratosphere included, it delivered a drop in temperatures of 1.8 to 2.6°F in Europe, but only a tenth as much globally—results remarkably close to likely events in the real world. The declining flows of ultraviolet radiation into the stratosphere triggered a slowdown in the westerly winds at ground level, says Shindell. That, in turn, caused winter cooling, particularly over land, in the higher latitudes of the Northern Hemisphere.

The stratosphere and its influence on polar and midlatitude winds thus seem to be a hidden amplifier that can turn small changes in solar radiation into larger changes in temperature in the polar regions of the planet. This is not the only amplifier in those regions. Ice and snow are important, along with the ocean conveyor and, maybe, methane. But it appears to be the critical ingredient that turns minor solar cycles into big climatic events. It makes sense of Bond's solar pulse and, perhaps, of tiny short-term variability in solar radiation.

Climate skeptics have sometimes argued that sunspot cycles correlate so well with warming in the twentieth century that greenhouse gases could be irrelevant. Mainstream climate scientists dismissed this idea because they could not see the mechanisms that might make this happen. The changes in solar radiation seemed much too small. Shindell's finding of a powerful stratospheric feedback to the solar signal have forced a rethink. But Shindell has not joined the climate skeptics. Far from it.

His conclusion is that for the first half of the century, the correlation between estimated solar output and Earth's temperature is not bad. And the stratospheric feedback might show how the sun could have driven some warming early in the century, followed by a midcentury cooling that made some fear an oncoming ice age. But since then, there has been no change in the solar signal that could be amplified to explain the recent warming. During the final three decades of the twentieth century, average solar output, if anything, declined, while global temperatures—not just at high latitudes but almost everywhere—surged ahead at what was probably a record rate. So, Shindell says, "although solar variability does impact surface climate indirectly, it was almost certainly not responsible for most of the rapid global warming seen over the past three decades."

For that most recent period, he says, it is clear that rising concentrations of greenhouse gases are the primary driver. But besides producing a general global warming, they have generated changes in the stratosphere that have produced a specific positive feedback to warming in the polar regions and the midlatitudes. The positive feedback has manifested itself through the apparently natural Arctic Oscillation and the SAM—cycles that appear to have gone into overdrive.

Only a fool would conclude from this that we don't need to worry so much about man-made climate change. On the contrary, Shindell's dramatic discovery of the stratospheric feedback suggests that the natural processes of temperature amplification are much stronger than those in most existing climate models. His newly discovered feedback seems set to continue, driving up temperatures in Arctic regions beyond the levels previously forecast. That additional warming is likely to unleash other feedbacks that will melt ice, raise sea levels, release greenhouse gases trapped in permafrost and beneath the ocean bed, and perhaps cause trouble for the ocean conveyor.

Relieved? I don't think so.

CONCLUSION: ANOTHER PLANET

Over the past 100,000 years, there have been only two generally stable periods of climate, according to Richard Alley. The first was "when the ice sheets were biggest and the world was coldest," he says. "The second is the period we are living in now." For most of the rest of the time, there has been "a crazily jumping climate." And now, after many generations of experiencing global climatic stability, human society seems in imminent danger of returning to a world of crazy jumps. We really have no idea what it will be like, or how we will cope. There is still a chance that the jumps won't materialize, and that instead the world will warm gradually, even benignly. But the odds are against it. There are numerous feedbacks— waking monsters, in Chris Rapley's words—waiting to provide the crazy jumps. Climatically, we are entering terra incognita.

The current generation of inhabitants of this planet is in all probability the last generation that can rely on anything close to a stable global climate in which to conduct its affairs. Jim Hansen gives us just a decade to change our ways. Beyond that, he says, the last thing we can anticipate is what economists call "business as usual." It will be anything but. "Business as usual will produce basically another planet," says Hansen. "How else can you describe climate change in which the Arctic becomes an open lake in the summer, and most land areas experience average climatic conditions not experienced before in even the most extreme years?"

I am sorry if you have got this far hoping for a definitive prognosis for our planet. Right now, the only such prognosis is uncertainty. The Earth system seems chaotic, with the potential to head off in many different directions. If there is order, we don't yet know where it lies. No scenario has the ring of certainty. No part of the planet has yet been identified as hold-ing an exclusive key to our future. No feedback is predestined to prevail. On past evidence, some areas may continue to matter more than others. But "the story of abrupt climate change will become more complicated be-fore it is finished," as Alley puts it. "We have to go looking for dangerous thresholds, wherever they may be."

For now, we have checklists of concerns. Melting Arctic ice, whether at sea or on land, could have huge impacts, both by raising sea levels and by amplifying global warming. Glaciological "monsters" could be lurking in Pine Island Bay or the Totten glacier. The whole West Antarctic ice sheet could just fall apart one day. El Niño may get stuck on or off, triggering megadroughts or superhurricanes. The Amazon rainforest may be close to disappearing in a rage of drought and fire that would impact weather sys-tems around the world. The oceans may turn into a giant lifeless acid bath. Smog may cripple the hydroxyl cleaning service or shut down the Asian monsoon. And the stratosphere may contain yet more surprises.

Methane is always lurking in the background, ready to repeat the great fart of 55 million years ago, if we allow it out of its various lairs. And the North Atlantic seems to hold a particular fascination. I keep coming back to Alley's disturbingly simple choice for the Gulf Stream as it surges north: sink or freeze? And to Peter Wadhams's lonely chimney, stuck out off Greenland somewhere northeast of Scoresby Sound, endlessly delivering water to the ocean floor. Until it stops. Who knows when? And who knows what will follow?

Quite a lot of this book has been taken up with climatic history. This is deliberate. The past shows more clearly than any computer model how the climate system works. It works not, generally, through gradual change but through periods of stability broken by sudden drunken lurches. And the past operation of the climate system reveals in their fully conscious state the monsters we may be in danger of waking.

But past climate does not provide a blueprint for the future. There are no easy analogues out there. We have already strayed too far from the tracks created by Bond's solar cycles and the other natural oscillations of the Earth system. Greenhouse gas concentrations are already probably at their highest level in millions of years; temperatures will soon join them. But the distinctive nature of our predicament goes a long way beyond that. Give or take the occasional asteroid impact, past changes have almost all been driven by changes in solar radiation, beamed down to us through the stratosphere. Earthly feedbacks such as biological pumps and spreading ice sheets, and threshold changes to marine currents and terrestrial vegetation, followed on the solar signal. This time, we are starting from the ground up, with a bonfire of fossil fuels that has shaken the carbon cycle to its core. Not only that: we are simultaneously filling the atmosphere with aerosols and assaulting key planetary features like the rainforests and the ozone layer. There can be no certainty about how the monsters of the Earth system will respond. We can still learn from the past, but we cannot expect the past to repeat itself.

When I first wrote at length about climate change, back in 1989, in a book called *Turning Up the Heat,* I warned that we passengers on Spaceship Earth could no longer sit back for the ride. We needed to get hold of the controls or risk disaster. But it was at heart an optimistic book. I figured that if *Homo sapiens* had come through the last ice age as a mere novice on the planet, then we could make it this time, too. We had the technology; and the economics of solving the problems wouldn't be crippling. I compared the task to getting rid of the old London pea-soupers of half a century ago. Once the decision was taken to act, the delivery would be relatively easy. We'd soon be wondering why we had dawdled for so long.

Fifteen years on, the urgency of the climate crisis is much clearer, even if the story has grown a little more complicated. But we are showing no signs yet of acting on the scale necessary. The technology is still straightforward, and the economics is only easier, but we can't get the politics right. Even at this late hour, I do believe we have it in our power to set Spaceship Earth back on the right course. But time is short. The ship is already starting to spin out of control. We may soon lose all chance of grabbing the wheel.

Humanity faces a genuinely new situation. It is not an environmental crisis in the accepted sense. It is a crisis for the entire life-support system of our civilization and our species. During the past 10,000 years, since the close of the last ice age, human civilizations have plundered and destroyed their local environments, wrecking the natural fecundity of sizable areas of the planet. Nevertheless, the planet's life-support system as a whole has until now remained stable. As one civilization fell, another rose. But the rules of the game have changed. In the Anthropocene, human influences on planetary systems are global and pervasive.

In the past, if we got things wrong and wrecked our environment, we could pack up and move somewhere else. Migration has always been one of our species' great survival strategies. Now we have nowhere else to go. No new frontier. We have only one atmosphere; only one planet.

APPENDIX: THE TRILLION-TON CHALLENGE

All the world's governments are committed to preventing "dangerous" climate change. They made that pledge at the Earth Summit in Rio de Janeiro in 1992. (The signatories included the U.S. and Australia, which both refused to ratify the subsequent Kyoto Protocol and its national targets for emissions reductions.) But what constitutes dangerous climate change? And how, in practice, can we prevent it?

For some people, dangerous climate change is already a reality. Many victims of recent hurricanes, floods, and droughts blame climate change. Such claims are usually impossible to prove. But that doesn't mean that our weather is not changing, says Myles Allen, of Oxford University. In essence, climate change is loading the dice in favor of weird and dangerous weather. "The danger zone is not something we are going to reach in the middle of this century," Allen says. "We are in it now." The 35,000 Europeans who died in the heat wave of 2003 were victims of an event that would almost certainly not have happened without the insidious increase in background temperatures that turned a warm summer into a killer.

But, despite such local disasters, most would argue that the critical aim in the quest to prevent dangerous climate change is to avoid crossing thresholds in the climate system where irreversible global changes occur—especially changes that themselves trigger further warming. There is no certainty about where such "tipping points" lie. But there is a growing consensus, especially in Europe, that the world should try to prevent global average temperatures from rising by more than 3.6°F above pre-industrial levels, or about 2.5 degrees above current levels.

Unfortunately, there is no certainty either about what limits on greenhouse gases will achieve that temperature target. We don't yet know how sensitive the climate system is. Current estimates suggest that to stack the odds in favor of staying below a 3.6-degree warming, we probably need to keep concentrations of man-made greenhouse gases below the heating equivalent of 450 parts per million of carbon dioxide. In practice, that probably means keeping carbon dioxide levels themselves below about 400 ppm. Let's call this the "safety-first" option.

Forgive me if I now abandon this language of parts per million. I find it an irritating and unnecessary abstraction. It seems to me much more sensible to talk in terms of tons of carbon instead. Then we can establish how much there is in the atmosphere and see more clearly how much we can afford to add before the climate goes pear-shaped.

The simple figures are these. At the depths of the last ice age, there were about 440 billion tons of carbon dioxide in the atmosphere. As the ice age closed, some 220 billion tons switched from the oceans to the atmosphere, raising the level there to about 660 billion tons. That's where things rested at the start of the Industrial Revolution, when humans began the large-scale burning of carbon fuels. Today, after a couple of centuries of rising emissions, we have added another 220 billion tons to the atmospheric burden, making it about 880 billion tons. If we want to keep below the safety-first concentration, we have to keep below 935 billion tons. So we have only about another 55 billion tons to go.

Currently, we pour about 8.2 billion tons of carbon into the atmosphere annually. Of this, a bit over 40 percent is quickly taken up by the oceans and by vegetation on land. The rest stays in the air, where its life expectancy is more than a century. So, for practical purposes, we are adding about 4.4 billion tons of carbon dioxide a year to the atmosphere. Even at current rates of emissions, that means that we will be above our 935-billion-ton safety-first target before 2020; and assuming that emissions continue to rise at the current rate, we will be there in less than a decade. Frankly, barring some global economic meltdown, there is now very little prospect of not exceeding 935 billion tons. If we had acted quickly after 1992, we could have done it. But the world failed.

If we are lucky—if climate sensitivity turns out to be a little lower than the gloomier predictions suggest—the 3.6-degree target may still be achieved while we allow carbon dioxide levels to rise significantly above 935 billion tons. We cannot be sure. There is already about 1 degree of warming "in the pipeline" that we can no longer prevent. But if we are feeling lucky—and with a nod to both round numbers and political reality—we might allow ourselves a ceiling of a trillion tons. Some would call that a "realistic" target, though others would brand it a foolish bet on a climate system we know little about.

The "trillion-ton challenge" is still a tough call. Literally, whatever target we set will require drastic cuts in emissions. Nature will probably continue to remove a certain amount of our emissions. But experts on the carbon cycle say that we must reduce emissions to around a quarter of today's levels before nature can remove what we add each year. Only then will atmospheric levels stabilize; only then will climate start to stabilize. The quicker we can do it, the lower the level at which carbon concentrations in the air will flatten out. Reaching the safety-first target of 935 billion tons of carbon dioxide would require an immediate and dramatic ditching of business as usual in the energy industry worldwide. Global emissions would need to peak within five years or so, to fall by at least 50 percent within the next half century, and to carry on down after that. A trillion-ton target could be achieved with more modest early cuts and greater reductions later.

Another consideration is the danger posed by the sheer speed of warming. Many climate scientists say that rapid warming may be more destabilizing to vulnerable systems like carbon stores and ice caps than slower warming. For this reason, it could be important to take some urgent steps to limit short-term warming while we get carbon dioxide emissions under control. And there is a way to do that—through a concerted assault on emissions of gases other than carbon dioxide that have a big short-term "hit" on climate.

Let me explain. Different greenhouse gases have different lifetimes in the atmosphere, ranging from thousands of years to less than a decade. For convenience, climate scientists usually assess their warming impact as if it operated over a century—carbon dioxide's average lifetime in the atmosphere. But this is rather arbitrary. And it has the effect of "tuning" the calculations to make carbon dioxide seem more important, and other gases less so. Most significant here is methane, which, however you measure it, is the second most important man-made greenhouse gas after carbon dioxide. Measured over a century, the warming caused by a molecule of methane is about twenty times as great as that caused by a molecule of carbon dioxide. But methane does most of its warming in the first decade, its typical lifetime in the atmosphere. It has a quick hit. Measured over the first decade after its release, a molecule of methane causes a hundred times as much warming as a molecule of carbon dioxide.

By following the scientists' conventional time frame, Kyoto Protocol emissions targets have underplayed the potential short-term benefits of tackling methane emissions. It is unlikely that the politicians who signed the protocol were even aware of this.

But underplaying the benefits has had an important effect on policy priorities. To take one example, if the British government decided today to eliminate all methane emissions from landfill sites, it would meet only a fraction of the country's Kyoto targets, because the Kyoto rules measure the impact of foregone emissions over the whole of the coming century. If the initiative were measured instead on its impact over the first decade, the benefits would be five times as great. The methane specialist Euan Nisbet, of London's Royal Holloway College, reckons that the short-term hit would be almost as great as banning all cars on the streets of Britain. And, if the rules had been drawn up differently, it would have been enough to entirely meet Britain's Kyoto target.

If the world is mainly concerned about the effect of greenhouse gases in fifty to a hundred years' time, then we should probably stick with the existing formula. But if we are also concerned about quickly reducing global warming to stave off more immediate disaster, then there is a strong case for coming down hard on methane now—on leaks from landfills, gas pipelines, coal mines, the guts of ruminants, and much else. "Cutting carbon dioxide emissions is essential, but we have neglected methane and the near-term benefits [acting on] it could bring," says Nisbet. He wants the Kyoto Protocol rules narrowed to a twenty-year time horizon. Jim Hansen takes a similar view. "It makes a lot of sense to try to reduce methane, because in some ways it's easier," he says.

Hansen also advocates action on soot, which he calculates to be the third biggest man-made heating force in the atmosphere. Soot, as we saw in Chapter 18, has a local cooling effect but a wider and more considerable warming effect. It sticks around in the atmosphere for only a few days, but while it is there, its effects are large. Action against soot and methane would not stop global warming. But it would give the world time to introduce measures against the chief culprit: carbon dioxide.

KYOTO POLITICS

The Kyoto Protocol, signed in 1997, was the first, tentative step toward implementing the Rio pledge to prevent dangerous climate change. Some forty industrialized nations promised to make cuts in their emissions of six greenhouse gases, including the "big two": carbon dioxide and methane. Different countries accepted different targets, and the countries of the European Union later internally reallocated theirs. Those cuts averaged about 5 percent, measured between 1990 and the first "compliance period," which runs from 2008 to 2012. The protocol included various "flexibility mechanisms" aimed at making more effective use of cleanup investment funds. They allow countries to offset emissions by investing in cleanup technology abroad and in planting trees to soak up carbon dioxide from the air, and to trade directly in pollution permits.

The protocol did not impose targets on developing countries, because their emissions per resident are mostly much lower than those of the rich industrialized world (some conspicuous exceptions include South Korea, Singapore, and several oil-rich Gulf states). The U.S. and Australia originally signed up to Kyoto targets, but then pulled out. The protocol came into force in 2005, and at the end of that year, its signatories agreed to start negotiations on tougher cuts to come into force after 2012.

So far, so good. But the current Kyoto targets are very small compared with the cuts in emissions that will eventually be needed. And the delay has effectively shut off the option of a safety-first limit on carbon concentrations in the atmosphere. Some European countries have set themselves informal targets of a 60 percent emissions reduction by midcentury, which is closer to what is needed. But even if all the Kyoto nations did likewise, they are responsible for only a minority of emissions today. So more cuts by other nations would still be needed.

Eventually, if the climate regime develops as many hope, every country and every major energy and manufacturing company will need a license to emit greenhouse gases. The system, some say, could even be extended to individuals. If we are to stop dangerous climate change, the number of licenses available will have to be very limited. So the question of how they should be shared out becomes critical. It is political dynamite. The very suggestion sets the industrialized and developing worlds at loggerheads. This is partly because the industrialized countries of Europe and North America have already used up something like half of the atmospheric "space" available for emissions, and partly because developing nations are coming under pressure to reduce their emissions before they have had a chance to industrialize.

Big developing nations like China and India may have high national emissions. But measured in ratio to population, their emissions remain low. While the U.S. and Australia emit around 5.5 tons of carbon a year for every citizen, and European countries average around 3 tons, China is still around 1 ton, and India below half a ton. Developing countries feel they are being asked to forego economic development to help clean up a mess they did not create. On the other hand, they increasingly see that climate change threatens their prospects for economic development. The only solution is to institute a rationing system for pollution entitlements, based on a shared view of fairness.

Perhaps the simplest blueprint is "contraction and convergence." Developed by a small British group called the Global Commons Institute, it is attracting support around the world. The contraction half of the formula would establish a rolling program of annual targets for global emissions. The targets would begin roughly where we are today, and would fall over the coming decades. They would be set so as to ensure that the atmosphere never passed whatever limit on carbon dioxide concentrations the world chose.

The convergence half of the formula would share out those allowable global emissions each year according to population size. So national targets might begin at about 1 ton of carbon per person and then fall to maybe half a ton by 2050 and to that much less again by 2100, depending on the global target chosen. Of course, at the start that would leave rich nations with too few permits and many poor nations with more than they needed.

So they would trade. The costs of buying and selling pollution licenses would be a powerful incentive for a global cleanup.

Fantasy politics? Maybe. But something on this scale will be needed if we are to prevent climatic disaster. And if the rich world wants the poor world to help clean up its mess, and save us all from dangerous climate change, then some such formula will be needed.

TECHNOLOGICAL FIXES

Politics aside, what are the practicalities of stabilizing climate? President George W. Bush may have become a pariah in environmental circles for refusing to sign the Kyoto Protocol, but he is right on one thing: ultimately, it will be technologies, rather than politics, that solve the problem. The only question is what politics will best deliver the technologies that will allow us to "decarbonize" the world energy system. Those technologies fall into four categories: much more efficient use of energy; a switch to low-carbon and carbon-free fuels; capturing and storing or recycling some of the emissions that cannot be prevented; and finding new methods of storing energy, such as hydrogen fuel cells.

The task sounds daunting. But, in truth, much of it goes with the grain of recent economic and industrial development. In the past thirty years, global carbon dioxide emissions have grown only half as fast as the global economy—thanks mostly to improved energy efficiency. And many of the new energy technologies we will need are already in use, offering benefits such as cheaper or more secure energy. The replacement of coal with lower-carbon natural gas, oil with ethanol made from biofuels, the development of wind and solar power, the proposed expansion of nuclear energy, and investment in energy efficiency all fall into this category. What is needed first is faster progress in a direction in which we are already headed.

The top priority should be energy efficiency. More than half of the immediate cheap potential for reducing carbon dioxide emissions lies in improving energy efficiency in buildings, transport, and industry. Much of it could be done at zero or even negative cost, because the cost savings would outweigh the investment. This is also the area where we as individuals can most easily make a difference—by buying energy-efficient light bulbs and appliances, insulating our homes properly, cutting down on car use, and choosing energy-efficient models such as hybrids.

Also in the short term, there is huge potential to equip the world's fossil-fuel-burning power stations with "scrubbers" to remove carbon dioxide and deliver it via pipelines for burial underground. The technology is already developed and only needs scaling up. The potential global storage capacity in old oil and gas wells alone approaches a trillion tons of carbon. The British government's chief scientist, David King, says that by 2020 Britain could be burying a quarter of its power-station carbon dioxide emissions in old oil fields beneath the North Sea.

Other technologies will take more development before they become cost-effective on a large scale. These include solar power, which is available but currently too expensive for widespread use, and turning hydrogen into the fuel of the future for transport. The idea here would be to manufacture hydrogen in vast quantities for use in batteries, known as fuel cells, to power cars. Hydrogen would become the "new oil." Hydrogen is manufactured by splitting water into hydrogen and oxygen, which is a very energy-intensive process. So if the energy for splitting water were generated by burning fossil fuel, there would be little environmental gain; but if the energy came from renewables, such as solar or wind power, that would change everything.

The hydrogen fuel cell is not so much a new source of energy as a new way of storing energy. It could be the only way to make cars truly greenhouse-friendly. And it may turn out to be the best way of utilizing fickle renewable energy sources like wind and the sun. The big problem with these energy sources is that wind cannot be guaranteed to blow (nor the sun to shine) when the energy is needed. But if the energy is converted into hydrogen, it can be kept for future use.

So what, exactly, would it take to deploy all these technologies in order to bring climate change under control? The most ambitious attempt so far to produce a simple global blueprint comes from Robert Socolow, an engineer at Princeton University. He admits that when he checked out the plethora of options for cutting greenhouse gases, he was overwhelmed, and figured that most politicians and industrialists would be, too. So he decided to break the task down into a series of technological changes that would each cut global emissions of carbon dioxide by about 25 billion tons over the coming fifty years. He called them "wedges," because the impact of each would grow gradually, from nothing in the first year to a billion-ton emissions cut in the fiftieth year. They would each cut a "wedge" out of the graph of rising carbon dioxide emissions.

Socolow proposed more than a dozen possible wedges, but said that seven would be necessary to stabilize emissions at current levels. But we need to do more than that: we need to stabilize actual concentrations of greenhouse gases in the atmosphere, and that would require reducing emissions from their current 8.2 billion tons a year to around 2.2 billion tons. So I have adapted Socolow's blueprint to allow for that tougher target. We might choose the following twelve wedges, each of which could cut emissions by about 25 billion tons over the coming half century, and reduce global emissions from the projected 15.4 billion tons a year by 2060 to 2.2 billion tons:

- universally adopt efficient lighting and electrical appliances in homes and offices;
- double the energy efficiency of 2 billion cars;
- build compact urban areas served by efficient public transport, halving future car use;
- effect a fiftyfold worldwide expansion of wind power, equivalent to 2 million 1-megawatt turbines;
- effect a fiftyfold worldwide expansion in the use of biofuels for vehicles;
- embark on a global program of insulating buildings;
- cover an area of land the size of New Jersey (Socolow's home state) with solar panels;
- quadruple current electricity production from natural gas by converting coal-fired power stations;
- capture and store carbon dioxide from 1,600 gigawatts of natural gas power plants;
- halt global deforestation and plant an area of land the size of India with new forests;
- double nuclear power capacity;
- increase tenfold the global use of low-tillage farming methods to increase soil storage of carbon.

ECONOMICS OF THE GREENHOUSE

How much might all this cost? In 2001, a team of environmental economists assembled by the IPCC reviewed estimates for stabilizing atmospheric concentrations of carbon dioxide by 2100. They ranged from a low
of $200 billion to a high of $17 trillion—almost a hundred times as much.
It seems extraordinary that estimates could range so widely. But, when
these are boiled down to their basics, it appears that much of the difference
depends on whether the modelers assumed that the necessary technical and
social changes would "go with the flow" of future change, or that everything would have to be grafted onto a society and an economy heading fast
in a different direction.

Put simply, the high estimates guessed that, under business as usual,
rising wealth would produce and require almost equally fast rises in emissions from burning cheap carbon fuels. Diverting from that path would
thus require preventing emissions of trillions of tons of carbon using expensive technologies that would not otherwise have been developed. The
lower estimates assumed that the world was already slowly losing its ad-
diction to carbon fuels, and that all we would need to do is make the switch
faster. They also took a rather different view of technological development,
seeing it as molded by a range of economic incentives. In this version, governments could shape technological development by stimulating markets.
Once the process was under way, innovation would go into overdrive, and
prices would fall away.

Some of the people involved in the IPCC study were instinctively hostile to major efforts to cut carbon dioxide emissions. The Yale environmental economist William Nordhaus suggests that "a vague premonition
of potential disaster is insufficient grounds to plunge the world into de-
pression." But let us assume that the real costs will be toward the top end
of the range. Would their adoption really push the world into recession?

The veteran climate scientist Stephen Schneider, of Stanford University, redid the arithmetic in 2002, assuming it would cost $8 trillion to
stabilize carbon dioxide concentrations by 2100. He found that the same
economists who predict doomsday if we try to tackle climate change also
believe that citizens of the world will be, on average, five times richer in a

hundred years than they are today. So he took the economists at their word and asked: How much would the $8 trillion bill for halting climate change delay those riches? The answer was just two years.

"The wild rhetoric about enslaving the poor and bankrupting the economy to do climate policy is fallacious, even if one accepts the conventional economic models," he told me when his analysis was published. Coincidentally, that was the week that Australia's prime minister, John Howard, announced that his country would not ratify the Kyoto Protocol because it would "cost jobs and damage our industry." Poppycock, said Schneider. "To be five times richer in 2100 versus 2102 would hardly be noticed." It was a small price to pay.

A small price to pay for what? What would we be buying with this trillion-dollar investment in a stable climate? That, of course, is impossible to answer, because we don't know the extent of what would be avoided. But we can easily see the scale of things, even today. Evidence of the cost of extreme weather is everywhere. The 1998 El Niño cost Asia at least $20 billion. Insured losses from extreme weather in 2004 hit a record $55 billion, which was promptly exceeded by an estimated $70 billion for 2005. Total economic losses for 2005, including uninsured losses, are expected to be three times higher: cleaning up after Hurricane Katrina alone may eventually cost $100 billion. Incidentally, a simple extrapolation of trends in insurance claims stemming from extreme weather in recent years suggests that they will exceed total global economic activity by 2060. That may be slightly wacky math, but it is sobering nonetheless.

Not surprisingly, economists disagree about the cost of inaction on climate change as much as they do about the cost of action. Some have attempted to assess the "social cost" of every ton of carbon put into the air. One recent review found a range from approaching $1,700 per ton down to zero. The British government, which commissioned the review, settled on a figure of $70 per ton. One reason for the wide range is accounting practices. Economists routinely apply a discount to the cost of anything that has to be paid for in the future. Dealing with climate change that may happen decades or even centuries ahead allows for huge discounts. Some economists say that very long-term impacts—such as the rise of sea levels as ice caps melt—should be discounted to zero.

This discounting of the future may be a convenient device for corporations, or even governments in their day-to-day business. But it is less clear how sensible it is for the management of a planet. If corporate finances or a nation's economy go wrong, shareholders can sell their shares and governments can print money or go cap in hand to the International Monetary Fund. But the planet, our only planet, is rather different.

Moreover, the existing estimates of social cost are based on IPCC studies that so far have not included many of the irreversible positive feedbacks to climate change that this book has concentrated on. So nobody has yet even asked what price should be attached to a century-long drought in the American West, or an enfeebled Asian monsoon, or a permanent El Niño in the Pacific, or a shutdown of the ocean conveyor, or the acidification of the oceans, or a methane belch from the ocean depths, or a collapse of the West Antarctic ice sheet, or sea levels rising by half a yard in a decade. Though, on reflection, these are perhaps questions best not answered by accountants.

GLOSSARY

Aerosols Any of a range of particles in the air, including soot, dust, and sulfates, that can intercept solar energy, sometimes scattering it and sometimes absorbing and reradiating it. Under different circumstances, they can either warm or cool the ground beneath and the air around.

African Humid Period The period after the close of the last ice age and before about 5,500 years ago, characterized by wet conditions in Africa, notably in the Sahara.

Albedo A measure of the reflectivity of a surface.

Anthropocene A new term to describe the past two centuries or so, during which human activities are seen to have dominated some key planetary processes such as the carbon cycle.

Arctic Oscillation A climate oscillation that occurs on timescales from days to decades. Measured by differences in air pressure between polar and nonpolar areas, and manifested in changing wind patterns that alter temperature. Related to (and sometimes synonymous with) the North Atlantic Oscillation.

Biological pump The process by which living organisms in the ocean draw carbon dioxide out of the atmosphere as they grow, and then deposit carbon on the ocean floor following their death. Has the effect of moderating the accumulation of CO_2 in the atmosphere.

Biosphere That part of Earth's surface, atmosphere, and oceans that is inhabited by living things.

Carbon dioxide fertilization effect What happens when higher concentrations of carbon dioxide in the air "fertilize" the faster growth of plants or other organisms.

Carbon cycle The natural exchange of carbon between the atmosphere, oceans, and Earth's surface. Carbon may be dissolved in the oceans, absorbed within living organisms and soils, or float in the air as carbon dioxide.

Carbon sink Anything that absorbs carbon dioxide from the air. Anything that releases carbon dioxide is a carbon source.

Chimneys A term coined by Peter Wadhams for giant whirlpools in the far North Atlantic that take dense water to the seabed. The start of the ocean conveyor.

Climate model A normally computerized simulation of the workings of the atmosphere. Often used to predict the effect of future changes such as an accumulation of greenhouse gases.

El Niño A periodic switch in the ocean currents and winds in the equatorial Pacific Ocean. A major perturbation in the global climate system.

Feedback Any by-product of an event that has a subsequent effect on that event. A positive feedback amplifies the original event, while a negative feedback dampens it. Key climate feedbacks include ice, water vapor, and changes to the carbon cycle. See also ice-albedo feedback.

Fossil fuel A fuel made from fossilized carbon, the remains of ancient vegetation. Includes coal, oil, and natural gas.

Gaia The idea, developed by James Lovelock, that Earth and its living organisms act in consort, like a single organism, to regulate the environment of the planet, including atmospheric chemistry and temperature.

Global warming Synonym for the greenhouse effect and climate change.

Greenhouse gas Any one of several gases, including water vapor, carbon dioxide, and methane, that trap heat in the lower atmosphere.

Gulf Stream The tropical ocean current that keeps Europe warm, especially in winter. Part of the ocean conveyor, and may be turned off at times, such as during ice ages.

Holocene The geological era since the end of the last ice age. Sometimes regarded as recently succeeded by the Anthropocene.

Hydrological cycle The movement of water between the oceans, the atmosphere, and Earth's surface through processes such as evaporation, condensation, rainfall, and river flow.

Ice ages   Periods of several tens of thousands of years when ice sheets spread across the Northern Hemisphere and the planet cools. Believed to be triggered by Milankovitch cycles and amplified by positive feedbacks. Recent ice ages have occurred roughly every 100,000 years. The last ended 10,000 years ago.

Ice-albedo feedback   A positive feedback on air temperature caused by the presence or absence of highly reflective ice. Thus, during warming, ice melts and is replaced by a darker surface of ocean or land vegetation that absorbs more heat, amplifying the warming. The reverse happens when cooling causes ice to form.

Ice sheets   The largest expanses of ice on the planet. There are currently three: Greenland, West Antarctica, and East Antarctica.

Interglacials   Warm periods between ice ages.

Intergovernmental Panel on Climate Change (IPCC)   A panel of scientists appointed by the UN through national science agencies to report on the causes of, impacts on, and solutions to global warming.

Isotope   One of two or more atoms with the same atomic number but containing different numbers of neutrons. For example: oxygen-16 and oxygen-18. The ratio of the isotopes in the air or oceans can vary according to environmental conditions, but will be fixed when the isotopes are taken up by plants, or air bubbles are trapped in ice. Thus isotopic analysis of ocean sediments, ice cores, and other leftovers from the past can be a valuable way of reconstructing past temperatures and other conditions.

Kyoto Protocol   The 1987 agreement on climate change, whose provisions include cuts in emissions by most industrialized nations during the first compliance period, from 2008 to 2012. The U.S. and Australia subsequently pulled out.

Little ice age   The period from the fourteenth to the nineteenth century when parts of the Northern Hemisphere were cooler than today.

Medieval warm period   The period from the ninth to the thirteenth century when parts of the Northern Hemisphere were notably warm.

Methane clathrates   Crystalline lattices of ice that trap large volumes of methane. Usually found at low temperatures and high pressures beneath the ocean bed or in permafrost.

Milankovitch wobbles   Various wobbles in the orbit of Earth than can influence climate over timescales of thousands of years. Believed to be the trigger for ice ages. Named after the Serbian mathematician Milutin Milankovitch, but originally investigated by the forgotten Scottish amateur scientist James Croll.

Nuclear winter   The theory that in a nuclear war, there would be so many fires that smoke would blanket the planet, causing massive cooling.

Ocean conveyor   Global ocean circulation in which dense surface water falls to the ocean floor in the Arctic and near Antarctica, travels the oceans, and resurfaces about a thousand years later in the warm Gulf Stream of the Atlantic. Prone to switching on and off, and perhaps a major determinant of global climate.

Ozone hole   An extreme thinning of the ozone layer seen in recent decades. Found each spring over Antarctica, but potentially could occur over the Arctic, too. Caused when man-made "ozone-eating" chemicals accumulate in the ozone layer. The immediate trigger for ozone destruction is low temperatures and sunlight.

Ozone layer   The ozone within the lower stratosphere, which protects Earth from harmful ultraviolet radiation from the sun.

Permafrost   Permanently frozen soil and rock found in the tundra regions of Siberia, Canada, Antarctica, and some mountain regions. Can reach a depth of more than 1.2 miles.

Precession   One of the Milankovitch wobbles that affects the axis of Earth's rotation. Changes the season when Earth is closest to the sun. Implicated in some climate changes during the Holocene.

Rainforest   Forest that depends on frequent rainfall, but also generates rain by recycling water into the atmosphere from its leaves.

Southern Hemisphere annular mode (SAM)   The Antarctic equivalent of the Arctic Oscillation. Responsible for strong warming of the Antarctic Peninsula in recent decades.

Stratosphere   A layer of the atmosphere starting about 6 to 9 miles up. Home of the ozone layer. Greenhouse effect causes it to cool, but it may act to amplify warming in the troposphere beneath.

Thermal expansion The warming and resulting expansion of water in the oceans. Along with the melting of land ice, it is causing a worldwide rise in sea levels.

Troposphere The lowest layer of the atmosphere, occupying the 6 to 9 miles beneath the stratosphere. The area within which our weather occurs. Greenhouse effect causes it to warm.

Ultraviolet radiation Solar radiation with wavelengths shorter than light but longer than X-rays. Harmful to living organisms, which are largely protected from it on Earth by the ozone layer.

ACKNOWLEDGMENTS

Where to start? In my twenty years of reporting on climate change for *New Scientist* magazine and others, innumerable scientists (and not a few editors and fellow journalists) have helped me get things mostly right. To all of them, thanks. I hope this book brings their work together in a form that many of them will find enlightening.

My greatest debt is to the synthesizers within the scientific community—the people who have tried to see the whole picture and to put their work into what seems to me an ever more frightening context. Their names recur throughout this book. But those who have specially helped me in person include Jim Hansen, Paul Crutzen, Jim Lovelock, Wally Broecker, Peter Cox, Peter Wadhams, Mike Mann, Richard Lindzen, Will Steffen, Richard Alley, Lonnie Thompson, Terry Hughes, Jack Rieley, Sergei Kirpotin, Euan Nisbet, Peter Liss, Torben Christensen, Crispin Tickell, Richard Betts, Myles Allen, Meinrat Andreae, Tim Lenton, Chris Rapley, Peter deMenocal, Joe Farman, Gavin Schmidt, Keith Briffa, John Houghton, Dan Schrag, Bert Bolin, Jesse Ausubel, Drew Shindell, Stefan Rahmstorf, Mark Cane, Arie Issar, Hans Joachim Schellnhuber, and the late Charles Keeling and Gerard Bond.

One always gets ideas from fellow writers. So thanks, too, to John Gribbin, Mark Lynas, Bill Burroughs, Doug Macdougall, Mark Bowen, Jeremy Leggett, Gabrielle Walker, and two historians of the climate change debate, Gale Christianson and Spencer Weart, whose books I have referred to in preparing this work. Thanks also to the organizers of the Dahlem conferences for making me welcome at an important event; to Carl Petter Niesen, in Ny-Alesund; and to the many people who have helped turn a germ of an idea into a completed book, including my agent, Jessica Woollard, and the editors Susanna Wadeson and Sarah Emsley.

NOTES ON THE REFERENCES

This is a far from complete list of the sources used in writing this book. But it includes the main written sources as well as others, summarizing information that could be of use to readers.

PREFACE

Wadhams's work on chimneys appears at greatest length in "Convective Chimneys in the Greenland Sea: A Review of Recent Observations" (*Oceanography and Marine Biology: An Annual Review* 2004, vol. 42, p. 29–56) and also in *Geophysical Research Letters* 2002 (vol. 29, no. 10, p. 76). Wadhams also spoke with me at length. For more on William Scoresby, see my article "Hell with a Harpoon" in *New Scientist,* 18 May 2002.

INTRODUCTION

The proceedings of the British government's Dangerous Climate Change conference appear at www.stabilisation2005.com. The resulting book can also be found at www.defra.gov.uk/environment/climatechange/internat/dangerous-cc .htm. Hansen's address to the AGU in late 2005 is at: www.columbia.edu/~jeh1/ keeling_talk_and_slides.pdf. Three overviews on abrupt climate change are: Richard Alley's *Abrupt Climate Change: Inevitable Surprises* (National Academies Press, 2002), especially chapter four; "Abrupt Changes: The Achilles' Heels in the Earth System" by Steffen et al. in *Environment* (vol. 46, p. 9) and Rial et al., "Non-Linearities, Feedbacks and Critical Thresholds with the Earth's Climate System" (*Climate Change,* vol. 65, p. 11).

1. THE PIONEERS

The journal *Ambio* had a special issue on Svante Arrhenius and his legacy in 1997 (vol. 26, no 1). I wrote about him in *New Scientist* in "Land of the Midnight Sums," 25 January 2003. Other sources include Gale E. Christianson's book *Greenhouse: The 200-Year Story of Global Warming* (Constable, 1999), which is also good on Callendar and Keeling. Many useful obituaries of Keeling were posted on news Web sites following his death in June 2005—for instance in the *Daily Telegraph* (www.telegraph.co.uk/news/main.jhtml?xml=/news/2005/06/24/db2402

.xml). And a good personal description of his early work appears at: www.mlo
.noaa.gov/HISTORY/PUBLISH/20th%20anniv/co2.htm.

2. Turning Up the Heat

The British newspaper mentioned in the first paragraph is the *Daily Mail.* The
column, by Melanie Phillips, "Global Warming Fraud," can be read at her Web
site: www.melaniephillips.com/articles/archives/000255.html. Christianson covers much of the early history of researching greenhouse gases. Brindley's paper
on the planet's radiation balance is in *Nature,* vol. 410, p. 355. See also: www
.imperial.ac.uk/P2641.htm.

The definitive consensus overview of the science of climate change in 2001 is
provided by the *Third Assessment Report of the Intergovernmental Panel on Climate
Change* (www.ipcc.ch), which will be superseded during 2007 by the *Fourth Assessment.* However, the *Fourth Assessment* is already out of date. It only accepted evidence published in peer-reviewed literature by the summer of 2005—missing
much new evidence of tipping points in the climate system.

Sherwood's 2005 research appears in *Science* (vol. 309, p. 1556). Parker's work
on the urban heat island appears in *Nature* (vol. 432, p. 290). For references to
Mann's work see the notes for chapter 33. Lassen and Friis-Christensen's original 1991 paper was in Science, vol. 254, p. 698. Lindzen is better known as a
polemical and op-ed writer (for instance www.cato.org/pubs/regulation/reg15n2g
.html), but he does have a track record of interesting research, such as "Does the
Earth have an adaptive infrared iris?" *Bulletin of the American Meteorological Society,*
vol. 82, p. 417.

Pat Michaels is another media regular. His exposition of the paradigm problem appears in his diatribe on climate science *Meltdown: The Predictable Distortion
of Global Warming by Scientists, Politicians, and the Media* (Cato Institute, 2004). For
a vigorous attack on Michael Crichton's book *State of Fear,* read Jeremy Leggett in
New Scientist, 5 March 2005. Oreskes's review of the scientific literature on climate
change appeared in *Science,* vol. 306, p. 1686.

3. The Year

I visited Honduras after Hurricane Mitch for the Red Cross. I wrote up my findings at www.redcross.int/EN/mag/magazine2001_2/heating.html. First reports
on how exceptional 1998 was appeared the following year (see, for instance, www
.sciencedaily.com/releases/1999/03/990304052546.htm). This was underlined in
2001 in the *Third Assessment Report of the IPCC.*

4. The Anthropocene

The proceedings of the Dahlem conference, at which I was introduced to many of the topics discussed here, are published as Earth System Analysis for Sustainability, Schellnhuber et al., eds. (Dahlem University Press, 2004). Crutzen discussed his work at length in his Nobel lecture (http://nobelprize.org/nobel_prizes/chemistry/laureates/1995/crutzen-lecture.html). His discussion of the Anthropocene first appeared in print in 2000 in the newsletter of the International Geosphere-Biosphere Programme (IGBP), no. 41. I interviewed him for *New Scientist*: "High Flyer," 5 July 2003. Alley's report is *Abrupt Climate Change: Inevitable Surprises* (National Academies Press, 2002). Many of the remarks by Alley and Steffen come from my interviews with them in 2003 and 2005.

5. The Watchtower

The reportage follows a visit to Ny-Alesund in September 2005. Kim Holmen discusses its role as "a watchtower for human-induced climate change" in *Polar Science in Tromso* (Polarmiljosenteret, 2004). Kohler's mass balance study appears in *Polar Research* (vol. 22[2], p. 145). Dobson's story can be read at www.atm.ox .ac.uk/user/barnett/ozoneconference/dobson.htm. The Bear Island research appeared in *Environmental Pollution,* vol. 136, p. 419.

6. Ninety Degrees North

McCarthy revealed the ice-free North Pole at http://news.bbc.co.uk/1/hi/world/americas/888235.stm. Scamdos's work is being updated all the time and appears at: http://nsidc.org/. Polyakov's warm water pulse was reported in 2005 in *Geophysical Research Letters,* vol. 32, L17605, DOI: 10.1029/2005GL023740; available at www.agu.org. The statement by glaciologists on the transformed state of the Arctic appeared in *Eos* in August 2005 (vol. 86, p. 309).

7. On the Slippery Slope

Hansen's "slippery slope" essay appears in *Climate Change,* vol. 68, p. 269. His "dangerous anthropogenic interference" remarks appeared in a lecture of that name to the University of Iowa, available, with much else of interest, from his Web site at: www.columbia.edu/~jeh1/. Box's remarks, and those of Bromwich and Alley, are from interviews conducted in 2005. Zwally's research was published in 2002 in *Science* (vol. 297, p. 218). Data on movement of the Jakobshavn glacier appear in *Nature* (vol. 432, p. 608), and the new findings on Kangerdlugssuaq from measurements by Gordon Hamilton of the University of Maine on a Greenpeace cruise in 2005 can be read at: www.greenpeace.org.uk/climate/climate.cfm ?UCIDParam=20050721151314.

8. THE SHELF

The demise of Larsen B is described by Hulbe at http://web.pdx.edu/~chulbe/science/Larsen/larsen2002.html. Alley discusses mechanisms at http://igloo.gsfc.nasa.gov/wais/pastmeetings/abstractso4/Alley.htm. I learned more from interviews with scientists at the British Antarctic Survey, and from Rignot and others at a conference on the Antarctic ice held at the Royal Society in London in late 2005 (www.royalsoc.ac.uk/news.asp?year=&id=3831).

9. THE MERCER LEGACY

I heard the story of Mercer from Thompson during interviews at his lab in 2005, and in correspondence with Hughes. Mercer's 1978 paper is in *Nature* (vol. 271, p. 321), and Hughes's 1981 "weak underbelly" paper was in the *Journal of Glaciology*, vol. 27, p. 518. Pine Island Bay was a major talking point at the Royal Society conference mentioned above, along with the state of the Totten and Cook glaciers. Vaughan's initial findings first emerged at http://igloo.gsfc.nasa.gov/wais/pastmeetings/abstractso5/Vaughan.pdf. Davis's paper on the East Antarctic ice sheet appeared in *Science* (vol. 308, p. 1898).

10. RISING TIDES

The plight of the Carterets reached the world via the BBC. See: www.sidsnet.org/archive/climate-newswire/2000/0093.html. Plans to abandon the islands and Tuvalu were reported by Reuters on 24 November 2005. I interviewed Teuatabo for *New Scientist* in 2000 ("Turning Back the Tide," 12 February 2000). Hansen discussed the history of sea level rise in his December 2005 lecture: www.columbia.edu/~jeh1/keeling_talk_and_slides.pdf.

11. IN THE JUNGLE

Nepstad's drought experiment is discussed in *Science*, vol. 308, p. 346, and at http://earthobservatory.nasa.gov/Study/AmazonDrought/. His plans for an experimental burn are discussed at www.eurekalert.org/pub_releases/2005–07/whrc-whro71905.php. The 2005 Amazon drought was widely reported, see http://news.bbc.co.uk/1/hi/world/americas/4344310.stm, for example. The Hadley Centre predictions appear in its report *Stabilising Climate to Avoid Dangerous Climate Change*, published in January 2005. The report by Gedney and Valdes appears in *Geophysical Research Letters*, vol. 27, no. 19, p. 3053.

12. WILD FIRES OF BORNEO

I visited Palangkaraya for *The Guardian* newspaper shortly after the fires of 1997–98 and received firsthand reports from locals. See also reportage in *Nature* in 2004 (vol. 432, p. 144). Rieley's calculations of emissions from the fires appeared in *Nature* (420, p. 61). The U.S. research corroborating his findings appeared in *Science* (vol. 303, p. 73).

13. SINK TO SOURCE

Fan's explosive *Science* paper appeared in vol. 282, p. 442. Ciais's work for CarboEurope appeared in *Nature* (vol. 437, p. 529), while Angert's paper appeared in the *Proceedings of the National Academy of Sciences (PNAS)*, vol. 102 (31), p. 10823, and Zeng's findings were in *Geophysical Research Letters*, vol. 32, L22709, DOI: 10.1029/2005GL024607; available at www.agu.org. Lawrence's work on permafrost is publicized at: www.ucar.edu/news/releases/2005/permafrost.shtml and in *Geophysical Research Letters*, vol. 32, L24401, DOI: 10.1029/2005GL023172; available at www.agu.org. Peter Cox presented his findings at the Dangerous Climate Change conference (see the notes for the Introduction, above) and published them in *Geophysical Research Letters*, vol. 30, no. 19, p. 1479. I found Canadell's work at: www.esm.ucsb.edu/academics/courses/595PP-S/Readings/VulnerabGlobalC.pdf. Kirk's findings on British carbon appeared in *Nature*, vol. 437, p. 245.

14. THE DOOMSDAY DEVICE

My story on melting permafrost appeared in *New Scientist* on 11 August 2005. Kirpotin's findings had yet to find a peer-reviewed publication in English at press time, but a revised version of his translated Russian paper appears at: www.mindfully.org/Air/2005/Palsas-Climate-Changes11aug05.htm. His findings were corroborated by Ted Schuur a year later in *Nature* (vol. 443, p. 71). I learned of Larry Smith's findings in e-mail interviews. The report suggesting that all plants make methane appeared in *Nature*, vol. 439, p. 187. I interviewed Christensen extensively during my visit to Stordalen in late 2005. His publications include *Geophysical Research Letters*, vol. 31, L04501, DOI: 10.1029/2003GL018680; available at www.agu.org.

15. THE ACID BATH

The Royal Society's study, "Ocean Acidification Due to Increasing Atmospheric Carbon Dioxide," appeared in June 2005, and can be found at: www.royalsoc.ac.uk. Turley presented her findings at the Dangerous Climate Change conference. Orr reported in *Nature* (vol. 437, p. 681). Falkowski's paper appeared in *Science* (vol. 290, p. 291).

16. THE WINDS OF CHANGE

Kennett and Stott's 1991 paper appeared in *Nature* vol. 353, p. 225. Dickens has published for instance at *Geotimes,* November 2004, p. 18. Alan Judd's seabed explorations were written up by Joanna Marchant in *New Scientist* on 2 December 2000. Norman Cherkis's paper was presented at the American Geophysical Union Spring Meeting 1997. Mienert discussed the Storegga slide in *Marine and Petroleum Geology* (vol. 22, p. 1) and in *Oceanography* (vol. 17, p. 16). Some other material comes from unpublished research he gave me during interviews. Nisbet discussed methane releases in a paper in the *Philosophical Transactions of the Royal Society,* Maths. Phys. Eng. Sc., vol. 360, no. 1793, p. 581. And David Archer produced an inventory of methane clathrates in *Earth and Planetary Science Letters,* vol. 227, p. 185.

17. WHAT'S WATTS?

Hansen's work on this is synthesized in his paper "The Earth's Energy Imbalance: Confirmation and Implications," published in *Science* (vol. 308, p. 1431) and available at: www.columbia.edu/~jeh1/hansen_imbalance.pdf. Read about the Global Albedo Project at: www-c4.ucsd.edu/gap/. Chapin's findings on Arctic albedo were published in *Science* (vol. 310, p. 627), while Betts's findings are in *Nature,* vol. 408, p. 187.

18. CLOUDS FROM BOTH SIDES

The 2004 Exeter meeting was a closed session of IPCC scientists. I was the only outsider attending. But most of the findings have since been made public. Stainforth's work appeared in *Nature* (vol. 433, p. 403), as did Murphy's (vol. 430, p. 768). Likewise, I was the only journalist attending sessions of the 2003 Dahlem Conference (see chapter 4), where Crutzen and Cox made their first calculations about the parasol effect, later written up by Cox in *Nature* (vol. 435, p. 1187). Wielicki responded in e-mail interviews and outlined some issues in *Science* (vol. 295, p. 841). Schwartz's remarks were made in an interview coinciding with the publication of his paper in the *Journal of the Air and Waste Management Association* (vol. 54, p. 1). Hansen wrote about black soot in the *Journal of Geophysical Research,* vol. 110, D18104.

19. A BILLION FIRES

The INDOEX Web site is at: www-indoex.ucsd.edu/. Remanathan and Crutzen discussed its findings in 2002 in *Current Science,* vol. 83, p. 947. Dale Kaiser's work on dimming appeared in *Geophysical Research Letters,* vol. 29, no. 21, p. 2042. Hansen's ideas appear in *Science,* vol. 297, p. 2250.

20. Hydroxyl Holiday

Prinn gave his warning in *Science* in 1995 (vol. 269, p. 187) and returned to the issue in the *IGBP Newsletter* No. 43 in 2000, and in *Science* in 2001 (vol. 292, p. 1882). Madronich raised his fears in 1992 in *Geophysical Research Letters,* vol. 19, no. 23, p. 465. And also here, a year later: www.ciesin.org/docs/011–457/011–457.html. I wrote a somewhat fanciful doomsday scenario for hydroxyl in a *New Scientist* supplement in April 2001. It can be found at www.gsenet.org/library/04 chm/hydroxyl.php.

21. Goldilocks and the Three Planets

Read all about Snowball Earth in the book of that name by my former *New Scientist* colleague Gabrielle Walker (Bloomsbury, 2003). And more from Kirschvink at: http://pr.caltech.edu/media/Press_Releases/PR12723.html. Lovelock gave his Gaian interpretation of the planet's history in books such as *The Ages of Gaia* (W. W. Norton, 1995). His most recent book is *The Revenge of Gaia* (Allen Lane, 2006). I explored Retallack's ideas about "The Kingdoms of Gaia," in *New Scientist,* 10 June 2001.

22. The Big Freeze

The best read on the ice ages and Agassiz and the intriguing James Croll is in *Frozen Earth* by Doug Macdougall (University of California Press, 2004). Shackleton's groundbreaking paper appeared in 1976 in *Science* (vol. 194, p. 1121). I took Berrien Moore III's analysis of carbon movements from the *Global Change Newsletter* No. 40 (December 1999, p. 1).

23. The Ocean Conveyor

Broecker's writings on the conveyor are extensive. Some key early papers are in *Nature* in 1994 (vol. 372, p. 421), *Scientific American* in 1995 (vol. 273, p. 62) and *Science* in 1997, (vol. 278, p. 1582). I interviewed him in late 2005. Schlesinger's paper appears on the Web site of the Dangerous Climate Change conference, along with Challenor's. Ruth Curry's paper on the great salinity anomaly was in *Science,* vol. 308, p. 1772. And Bryden's paper appeared in *Nature,* vol. 438, p. 655.

24. An Arctic Flower

Alley splendidly describes the science of the Younger Dryas (and many other things) in his book *The Two-Mile Time Machine* (Princeton University Press, 2000). Read about how humans fared in William Burroughs' Climate Change in Prehistory (Cambridge University Press, 2005). The latest thinking on the emptying of Lake Agassiz is in *Eos*, vol. 86, p. 465. Chiang's paper appeared in *Climate Dynamics* (vol. 25, p. 477). Alley explored events 8200 years ago in *Quaternary Science Reviews*, vol. 24, p. 1123.

25. THE PULSE

The best study of the events of the Little Ice Age remains the book of that name by Jean Grove (Routledge, 1988). Bond's pioneering work on "the pulse" and its links to the era appeared in *Science* (vol. 278, p. 1257 and vol. 294, p. 2130). His work is summarized at: www.ldeo.columbia.edu/news/2005/07_11_05.htm. Read too Peter deMenocal's paper with Thomas Marchitto in *Geochemistry Geophysics Geosystems* (DOI: 10.1029/2003GC000598) and his essay "After Tomorrow" in *Orion*, Jan./Feb. 2005; plus Shindell's "Glaciers, Old Masters and Galileo" at: www.giss.nasa.gov/research/briefs/shindell_06/; and Christina Hulbe in *Paleoceanography* (vol. 19, PA1004).

26. THE FALL

Useful analysis of how the Sahara became a desert include Robert Kunzig's "Exit from Eden" in *Discovery*, January 2000, Claussen's paper in *Climate Change* (vol. 57, p. 99), and deMenocal in *Quaternary Science Reviews*, vol. 19, p. 347. Haarsma's theories are articulated in *Geophysical Research Letters*, vol. 32, L17702, DOI: 10 .1029/2005GL023232; available at www.agu.org. DeMenocal looks at megadroughts through the late Holocene in *Science* (vol. 292, p. 667); and Richard Seager's study is at www.ldeo.columbia.edu/res/div/ocp/drought.

27. SEESAW ACROSS THE OCEAN

The Bodele dust reservoir is discussed in *Nature* as "the dustiest place on Earth" (vol. 434, p. 816). I learned of Schellenhuber's ideas on links between the Sahara and the Amazon in conversations. They seem intuitively sensible but remain, so far as I know, unquantified.

28. TROPICAL HIGH

I interviewed Thompson at length about his career and ideas in 2005. There is also a highly readable book about him called *Thin Ice* by Mark Bowen (Henry Holt, 2005). Key publications include *Climatic Change*, vol. 59, p. 137, and *Quaternary Science Reviews*, vol. 19, p. 19. His Web site is at: www-bprc.mps.ohio-state.edu/ Icecore/GroupP.html#lonniethompson.

29. THE CURSE OF AKKAD

The story of Akkad and other tales of climate and civilization appear in *The Winds of Change* by Eugene Linden (Simon & Schuster, 2006). DeMenocal looks at the collapse of Akkad in *Geology*, vol. 28, p. 379. Weiss's original paper appeared in the *Journal of the American Oriental Society*, vol. 95, p. 534. Issar explores similar collapses in the Middle East at the time with Mattanyah Zohar in *Climate Change: Environment and Civilization in the Middle East* (Springer, 2004).

30. A CHUNK OF CORAL

I wrote about Dan Schrag's find and its implications for El Niño in *New Scientist*, 9 October 1999. He published his findings in *Geophysical Research Letters* (vol. 26, no. 20, p. 2139). El Niño has many chroniclers these days, including Richard Grove and John Chappell's *El Niño: History and Crisis* (White Horse Press, 2000) and El Niño in History by Cesar Caviedes (University Press of Florida, 2001). Rodbell's compelling paper is in *Science* (vol. 283, p. 516). Latif's modeling of El Niño's future appeared in *Nature* (vol. 398, p. 694). Read about the Peruvian potato farmers at www.columbia.edu/cu/pr/00/01/pleiades.html.

31. FEEDING ASIA

Mike Davis wrote passionately about the effects of El Niño and failed monsoons in the late nineteenth century in *Late Victorian Holocausts* (Verso, 2001). Overpeck's analysis of the monsoon's potentially troubled future appeared in *Nature*, vol. 421, p. 354. Analysis of the different interpretations of the links that sustain the monsoon emerged from conversations with Mark Cane, Broecker, Alley, Thompson, and others.

32. THE HEAT WAVE

The 2003 heat wave was summed up at: www.earth-policy.org/Updates/Update 29.htm. The link to global warming was articulated by Allen in *Nature* (vol. 432, p. 610). The study of Burgundy vineyards appeared in *Nature* (vol. 432, p. 289). Betts warned about the extra threat to cities in *PNAS* (DOI 10.1073/pnas .0400357101).

33. THE HOCKEY STICK

Read the IPCC summary for policymakers at: www.ipcc.ch/pub/spm22–01.pdf. Early versions of the hockey stick were discussed in *Nature* (vol. 392, p. 779) and *Geophysical Research Letters* (vol. 26, no. 6, p. 759). Other write-ups of Mann's work and the controversy it created were included in *Scientific American* (March 2005, p. 34) and *Mother Jones* (18 April 2005). McIntyre and McKitrick set out their case in 2003 in *Energy and Environment*, vol. 14, p. 751. Mann's side of the debate, with commentary from some critics, appears on a Web site run by him and others: www.realclimate.org. Recent scientific analyses of the debate include Osborn and Briffa in *Science* (vol. 311, p. 841).

34. HURRICANE SEASON

I spoke to Corky Perret for a feature in *New Scientist,* "Is Global Warming Making Hurricanes Stronger?" (3 December 2005). Webster's paper appeared in *Science* (vol. 309, p. 1844). Emmanuel first predicted a big increase in hurricane destruction in *Nature* in 1987 (vol. 326, p. 483). He was more sanguine when, with others, he reported in the *Bulletin of the American Meteorological Society* in 1998 (vol. 79, p. 19), but returned to the barricades in *Nature* in 2005 (vol. 436, p. 686). Trenberth made his warnings earlier that year in *Science* (vol. 308, p. 1753). Gray's efforts to refute these claims were not carried in the major journals, but can be seen at his Web site: http://typhoon.atmos.colostate.edu/. The story of "hurricane" Catarina is told at: www.met-office.gov.uk/sec2/sec2cyclone/catarina.html.

35. OZONE HOLES IN THE GREENHOUSE

Farman's landmark paper on the ozone hole appeared in *Nature* (vol. 315, p. 207). Crutzen discussed how lucky the world had been in his Nobel lecture (see chapter 4). Hormes's and Shindell's thoughts come from personal interviews in Ny-Alesund and New York, respectively. The mechanisms that might cause ozone depletion to produce rapid climate change were discussed by Hartmann et al. in *PNAS,* vol. 97, p. 1412.

36. THE DANCE

The debate between the polar and tropical schools has never been properly articulated in the journals, so this chapter is pieced together from interviews with the participants, many of them in New York. Goldstein's paper appears in *Science* (vol. 307, p. 1933). Crutzen's comments came from an interview I conducted.

37. NEW HORIZONS

Similarly, much of this chapter derives from conversations rather than written papers. The idea of a bipolar seesaw is discussed by David Sugden in *Planet Earth,* journal of the Natural Environment Research Council, in autumn 2005. Read about lakes beneath the Antarctic ice at: www.earth.columbia.edu/news/2006/story01-26-06.html. Recent changes to SAM are reviewed by King in *Geophysical Research Letters,* vol. 32, L19604, DOI: 10.1029/2005GL024042; available at www.agu.org. Shindell's key papers appear in *Science* (vol. 284, p. 305 and vol. 294, p. 2149) with useful summaries at www.giss.nasa.gov/research/news/20041006/ and www.giss.nasa.gov/research/briefs/shindell_04/.

Conclusion

My earlier book, *Turning Up the Heat,* long out of print, was published by The Bodley Head in 1989 and in paperback by Paladin later the same year.

Appendix

Much of what appears here was presented at the Dangerous Climate Change conference, whose proceedings can be found at www.stabilisation2005.com.

INDEX